The Weather

The Weather

The Truth about the
Health of Our Planet

ANTHONY SMITH

HUTCHINSON
LONDON

First published in the United Kingdom in 2000 by Hutchinson

The Random House Group Limited
20 Vauxhall Bridge Road, London SW1V 2SA

Random House Australia (Pty) Limited
20 Alfred Street, Milsons Point, Sydney,
New South Wales 2061, Australia

Random House New Zealand Limited
18 Poland Road, Glenfield,
Auckland 10, New Zealand

Random House (Pty) Limited
Endulini, 5a Jubilee Road, Parktown 2193, South Africa

The Random House Group Ltd Reg. No. 954009
www.randomhouse.co.uk

A CIP catalogue record for this book is available
from the British Library

Papers used by Random House
are natural, recyclable products made from wood grown in
sustainable forests. The manufacturing processes conform to
the environmental regulations of the country of origin

ISBN 0 09 180090 0

Typeset in Baskerville by MATS, Southend-on-Sea, Essex
Printed and bound in Great Britain by
Clays Ltd, St Ives plc

CONTENTS

A PROLONGED DEDICATION

To my father

He was obsessed by weather. Even in his earliest childhood diaries he would write 'Wind north-west, brisk, no rain' before any other kind of entry – the death of a pet, a sister's illness, the finding of a favoured moth. In retirement he continued with this dominant enthusiasm, and I loved to hear of it. There was I, sweating it out in Lagos (minimum recorded temperature 69 degrees Fahrenheit) or, worse still for heat, in the desert around Khartoum, when a letter would arrive from Dorset: 'Quite a sharp frost yesterday. Some ice by the greenhouse. Wind northerly.' Like cool drinks or air conditioning these reminders of another world served as balm for an overheated frame.

He was also mono-regional, if there is such a term. Nowhere else was of as much consequence as his home, his garden, his immediate area. A hurricane may have blown Florida to bits, or a three-year drought was bringing Ethiopia to its knees, but 'Light breeze from south-west; rain forecast, only few drops so far; chilly night' still persisted. External happenings, however awesome, were always of minor (or zero) interest. So were those from wherever I happened to be living. Sometimes I would mimic his weather enthusiasm, and even attempt to upstage his temperate events: 'Light rain yesterday, 12 inches in 24 hours. Ten killed by lightning. Mudslides everywhere.' There was never any response to such tropical rivalry, save further reports from deepest Dorset: 'Oak in leaf before ash, so we will get a splash. Perhaps. Only 0.92 inches so far this month. Gooseberry bushes need a good soaking. Winds light and variable.'

This dedication is being written partly because I do not think my father was unique. He never seemed to grasp that his precious homestead existed on a planet. He did not see his light winds and fearsome downpours – 'One month's average rainfall in one week. Garden sodden, with another 0.4 inches yesterday. Good grief!' – in their broader context. When meteorologists showed their charts on the television screens I swear that his eyes narrowed solely to the spot he knew so well. Was there a raindrop over his home? Or a cloud? Or a suggested temperature? As for the whirls of depressions or anti-cyclones out in the Atlantic, they were of no appeal. He did not live out there and, even if all those circles served as indicators of weather to come, they were too confusing and somehow unreal.

So how many of us, including my father, are unofficial members of a flat earth society, viewing our own particular world much as an ant must see its piece of territory? We know, in truth, that we live on a planet, but have not absorbed that fact. We refer to sunset and sunrise although the sun is doing nothing of the sort, merely vanishing because we are revolving. We say the moon is half-full or even full, as if the moon itself is somehow altering. (And I well remember, when writing of the lunar landings, hearing from a reader who chided the Americans for firing objects and even people at the moon when it was only half in evidence. 'Why don't they wait until it's bigger,' she proclaimed. I think my father would have sympathised.)

If we do truly comprehend that we on Earth are living on a planet I think we should know, just as we know when it is Christmas or if it is summertime, when our orbiting home is closest to the sun. All orbits are elliptical rather than precisely circular, and ours is no exception. It so happens that Earth is three million miles nearer the sun at its closest approach, and therefore receives more heat at that time than when it is furthest away. This extra heat, over 1% more, is of considerable consequence. It means that winters in the northern hemisphere are not as cold as they would be without this extra warming, it being a fact that our closest approach occurs during the northern wintertime. If some alien were describing planet Earth, and telling of its basic parameters to those back home, the details of its ellipse would be high, if not foremost, on that alien's list. So when (as I once asked my father, before receiving a prolonged stare, as if requesting that week's pop chart leader) is this date? He did not even wish to know, although – as a possible *aide-mémoire* – his first-born had arrived during that all important time at the start of January.

He did not respond to this piece of information by working out that planet Earth is receiving its least solar energy at the converse time of year, namely the beginning of July. School-teachers are familiar with the look they receive when they present youngsters with some riveting truths – about the Treaty of Utrecht, the War of Jenkin's Ear, or the Spanish succession – they might have quoted that day's fat-stock prices for all the excitement they arouse. So too when I spoke in Dorset about cold fronts, isobars, alto-cumulus, depressions gyrating clockwise in the south, or the beginning of July.

Putting my father to one side, as it were, I think there are quite a few of us who, like him – and to use another novel term – are northist. We see the North Pole as pre-eminent, and the South Pole as inferior by being at the bottom end. We also know, with all our northist prejudice, that summer occurs from June to August, more or less, but certainly not in January. There is no astronomical reason for this northern form of favouritism, save that so many early map-makers, mariners and savants in general came from the hemisphere of which Europe is a part. A football has no top or bottom, and neither does Earth. The planet is certainly rotating, and on an axis through both poles, but neither has some form of celestial advantage over the other. They are as equal as could be.

There is also no reason, save for historical tradition, why all the maps we draw have their more northern portions at the top. Had Australians or Tierra del Fuegians been first to portray the ground they inhabited in a two-dimensional fashion they could well have placed Antarctica at the top of all their charts. Florida, when that had been encountered, would then have been above Connecticut, and Cornwall over Scotland. Summer would also have been, in much of the world's most famous literature, associated with December and January. 'Oh, to be in Buenos Aires now that October's there.' In the *Oxford Dictionary of Quotations* there are many references to April, but not one which makes it an autumnal happening.

The need for flat maps helps us to forget we live upon a sphere. Great circles, when we encounter them, help us to remember. They indicate the shortest distance between two points and are extremely relevant, for example, to all aircraft flying from one continent to another. We passengers may think our pilot has lost his senses when, after leaving London for New York (with us knowing that New York is more southerly), he promptly steers to the north of our starting place. That ill-fated Pan American aircraft which exploded over Lockerbie in Scotland was almost 4 latitude degrees north of its take-off spot at the time of the disaster, although heading for a more southern region of eastern North America. I was once flying from Japan to Britain when the sun set over my left shoulder. So where would it rise, I wondered, when that night was done? It eventually rose over my left shoulder, leaving me with a headache of concern but everyone else unmoved.

My father, to bring him back again – as this is his dedication – was so interested in local weather, in local bird life, and in local everything that was seasonal, that no curiosity existed for anything beyond his immediate realm. He did not care why the hands of clocks go round clockwise, which they would not do if sundials had been invented south of the equator.

Instead he was more concerned with a batch of alleged truths, such as March's arrival 'like a lamb' certain to make its ending 'like a lion', or vice versa, and he was positively uninterested that hurricanes always fade in their fury whenever meeting land. Every day he examined his rain gauge to measure its contents, whether 0.12 inches or even a generous 0.4, but he never cared that Cherrapunji in northern India has been known to experience 75 feet of rain in what it calls, with reason, a wetter than normal year – and almost all of that falling in 6 months. His maximum and minimum thermometer, screwed to the outer porch, was inspected every morning to see if a cold snap, perhaps 6 Fahrenheit degrees of frost, had happened overnight. I knew he cared nothing for those in Verkhoyansk who experience minus 71 degrees Centigrade when their portion of Siberia is being particularly chilly, and when any foray out of doors demands eye-protectors if corneas are not to freeze causing blindness.

Alas, Dorset is no longer so well recorded, and thermometers and rain-gauges in one particular locality are no longer checked each day. I know this book of mine (for him) would not have been seized upon to make him more mindful of the planet on which he lived in such a minute part, but I like to think it might have done had he been spared the time.

Or perhaps he and his generation were born too soon. They did not fly everywhere, and along great circle routes, as is the modern way. They did not have instant pictures from every mishap around the world, or e-mail, or hand-held mobile phones for communication with anyone at any time. It is easier now for us to accept our planetary situation. The six billion of us have but one home. No longer can we consider that a part of Dorset, or the whole of England, or Europe, or the northern hemisphere, is somehow independent of the rest. There is this single sphere called Earth. It, rather than a piece of it, is an earth of majesty, an Eden, a demi-paradise, a fortress built by Nature for herself, a precious stone set in a silver sea; but there is the unhappy extra that we may have damaged it immensely. Our world is getting warmer; that is a fact. We may be in part to blame; that is a possibility. We must attend to our situation with greater care; that is a certainty.

Future fathers of future sons may well record in future diaries: 'Global temperature still rising, but less speedily. Greenhouse gases also slackening their acceleration, particularly CO_2. Sea level now constant. Glaciers beginning to form and advance again. Reforestation progressing actively. Hurricanes less devastating and less frequent. Anxiety felt at the turn of the 21st century now abating. World population actually dropping.'

Or, of course, they may write nothing of the sort.

INTRODUCTION

In 1970 I published a book called *The Seasons*. It dealt with this subject in its broadest sense, covering not only the basic astronomical situation which causes annual cycles here on Earth but all their major differences, at the poles, by the equator, within the temperate zones. It examined not only human lives in those areas but those of trees and plants and animals. There were sections on hibernation, on adaptive colouration, and on all the triggers to sexual development, as well as the various forms of internal rhythm, the biological clocks which serve as calendars for so many living things.

I thought the book wide-ranging. It seemed to cover every aspect of the subject – hurricanes, migration, cave existences, deserts, ice ages, photoperiodicity, jet lag, monsoons, tornadoes, floods, hail, equinoctial gales, lunar cycles, and certainly every form of weather in all its variation and unpredictability. The book no doubt fell short in general competence, in matching fulfilment with expectation, but at least its numerous chapter headings and its index showed the range of this amazing subject, whether or not the various portions of this material were dealt with satisfactorily.

From today's viewpoint that broad approach led to a most positive form of failure. The book assumed that everything seasonal was more or less as it had always been, with colder times and warmer times, windier years and wetter years, and variation occurring – as was only natural – in every aspect of a planetary life. There was talk of conservation in the late 1960s, with alarm bells sounding in various quarters, but these were particular rather than general. The livelihoods of whales and tigers, of people in desert regions, of animals and humans too near to each other – all such items were disturbing, but there was very little talk, if any, about the planet as a whole being amended.

As for the various single-issue cries for help, some solutions promptly surfaced: a moratorium on whaling, more wildlife parks, better and safer pesticides, more attention and money for the third world's needs, though by no means was every problem solved. Each one was seen almost in isolation: 'Save the tiger'; 'Fur coats look best on animals'; 'Ban Aldrin, Dieldrin and Heptachlor'; 'Stop the bloody whaling'. A modicum of such concern had always existed, but only in the 1960s did conservation suddenly become populist. After motherhood, said someone, it had become 'the safest thing to be for'.

Then, no less suddenly, and shrinking all other topics appropriately, there was talk of alteration to the planet itself. An awesome cry began to detail, as was said, the 'monstrous experiment' being conducted upon our only home. Even motherhood – and particularly excessive motherhood – fell from grace when confronted by talk of increasing UV radiation, of rising greenhouse gases, of excessive heat, of melting ice, of expanding deserts and every other harmful consequence of human activity. The new concern, the global concern, hit the headlines and then, no less acutely, our collective consciousness. 'What is the point of a decent home if you don't have a tolerable planet to put it on?' as Henry David Thoreau asked all those years ago. More than a century and a half were to pass before we truly heard his words, and slightly longer before we realised that our straightforward livelihoods were achieving the damage. We were doing it in no more awful a manner than by owning a car, advancing our lives, acquiring possessions, keeping our houses warm and our children fed. Was that not progress? Were we not supposed to better our existences, and those of our successors?

From today's standpoint it seems amazing that a book called *The Seasons*, published only 30 years ago and purportedly wide-ranging about weather, climate, natural hazard and human living, should be so replete with gaps and missing words. Nowhere within that volume – whatever the edition – was there any mention of global warming, ozone depletion, carbon dioxide quantities, excessive methane, glacial retreat, El Niño, La Niña, rising sea levels, low-lying inundation or greenhouse happenings. Natural disasters were fully described, but there was no suggestion that human activity might be, in part, to blame for an increase in their frequency or their severity.

In that book's index, ranging from adaptation and aestivation to zebras and zeitgebers, there was no entry to suggest that our world might itself be changing, and deleteriously, due to human error. Of course people could be blamed for the extinguishing of many species, for bringing others to the brink of existence, for felling entire forests, and for emptying the oceans of much desirable life, but the loss of tigers or blue whales did not affect the planet. Neither did the disappearance of cod nor the survival of more people. However, news of ozone loss and a gain in greenhouse gases did produce a new level of concern. There was even talk that the planet might become so much less tolerable a place on which to live that living itself might be jeopardised.

The book's omission of these frightening possibilities was not, I hasten to add, entirely the author's fault. When the book was being written, largely in the rustic peace of northern Wales, the shrinkage of the ozone layer (what on earth was that?) and the increase in its 'hole' (whatever and wherever that was) had not yet been reported. El Niño had been so named by fishermen in the 19th century, but its effects were presumed to be of concern solely to Peruvian fishermen and Peruvian birds. It therefore did not gain a mention in a book whose potential audience lived, almost entirely, in the northern hemisphere. The change in southern Pacific currents, leading to warmth in place of cold, was of modest universal interest until the 1980s and, in particular, the 1990s. Its effects were then realised to be so devastating and so influential, stretching far beyond Peru, that everyone on the planet – wherever they lived – learned the charming name for this most uncharming upset to the global scene. If only Peruvians were suffering, so be it; but if droughts in Australia, rain torrents in Africa and mud-slides in California were all related to, and possibly caused by, this south Pacific shift in water temperature, the world took note.

It had long been known that the consumption of fossil fuel could be harmful. Run a car in a closed garage, and anyone in that space will quickly be dead. Smogs were known to prevent the dissipation of smoke, and people died in consequence, but all concern about the burning of coal, of coke, and of oil in the years immediately after World War Two concentrated upon the exhaustion of fossil fuels rather than, to twist the phrase around, the exhausts which they produced. Smoke and smog were unpleasant, but there was not wholesale and global anxiety about so much effluent.

Concern about human interference with the weather was not then manifest. The major environmental conference at Stockholm had not yet occurred, let alone the further and similar international gatherings at Montreal, Rio de Janeiro, Kyoto and Buenos Aires. There were some early warnings from conservationists. Rachel Carson, for example, noted the lack of birds in springtime, although she did not suggest that spring's climate might itself be altering. There was worry about pollution and pesticides, loss of species and reduction in wilderness, and there was extreme unhappiness about nuclear contamination of the environment, with 50-megaton bombs – ye gods, some 2,500 times the power of Hiroshima's destroyer – being exploded in the atmosphere. There was yet more worry about nuclear war itself, but no overall concern that planet Earth was becoming less habitable merely by our collective wish to acquire, to have a better refrigerator, to speed along more highways, and to travel more energetically to – well, to everywhere.

Instead we were revelling in the consequences of a longer-lasting peace than had occurred in the 20th century's first half. During those first 50 years there had been great misery, wholesale unemployment, disabling recession, severe shortages, and two dreadful global wars killing tens of millions. The prolonged peace thereafter was not entirely global, as other wars broke out and millions more were killed, but the years post-World War Two were, in general, of plenty rather than penury, of more money to spend, of national freedoms, and of more consumers – by 90 million a year – than ever before. 'Enjoy the war, the peace will be terrible,' proclaimed graffiti when Nazi Germany was being overrun. 'Enjoy the peace as wars were terrible,' then formed a kind of slogan for the new consumer age, as people consumed more of everything than had ever been known before. Hence the disarming jolt when a new truth emerged that our actual home, our planet, our single dwelling place, was being jeopardised by so much peace, so much progress, and so much fulfilment of a general aspiration.

This book, therefore, is *The Seasons Mark II*. It still defines the planetary condition, with its solstices, equinoxes, gyrations and perturbations, as that is the basic raw material of any orbiting body. It also outlines the calendar by which we measure the various cycles that are involved, the days and nights, the summers and winters, the rainy times, the drier times, the rhythms fundamental to our lives. As for the tropics, the temperate areas, and both arctic circles, these are still as they were. So too the mountains, oceans, continents and every other major feature of our Earth. We, with all our different forms of livelihood, are as influenced by climate as we ever were. Therefore climate in all its forms is determinedly addressed, as are animal, plant and human adaptation to all its vagaries. Change can only be understood, and realised and (perhaps) feared, if the parameters lying at its base are solidly absorbed.

Living on a planet, whether that planet is being abused or not, is difficult to accept. One aim of this book – as with its predecessor – is to create a better awareness of that simple fact. Why do we have the fundamentals of summer and winter? Why do hurricanes and jet streams occur particularly in definite seasons, and why do tornadoes tend to travel north-east (in the northern hemisphere) although hurricanes move north-west? Why are southern winters colder, and why do winds travel anticlockwise around depressions in the northern hemisphere (and vice versa in the south)? What are doldrums? How thick is the atmosphere which gives us all our life? How is it that birds lay their eggs in spring, an age-old event we take so much for granted that we scarcely stop to wonder at its annual miracle? Did 2,000 children really die in London's 1911 heat wave? Do giraffes get struck by lightning more than other animals? Why is there no life on Mars or Venus but such an abundance of it here?

Only then, having absorbed the framework of life on planet Earth, can the possibility of man-made change make more sense. If there is more carbon dioxide in the atmosphere will that inevitably create more heat? If more heat is a consequence will there not be more evaporation from the oceans, and therefore much more cloud? If there is more cloud will that not reduce the quantity of solar energy actually reaching Earth, therefore creating global cooling instead of warmth? And if a volcano erupts most devastatingly, as Pinatubo did in 1991, does that not create more effluent, and initiate more climate change, than all of mankind's effulgent chimneys and exhausts? Are natural forces not so tremendous that the power of human beings, for all their wrongful labour, is puny by comparison?

There have always been colder times and warmer times, but the trick is to know whether any modern alterations have us as cause. The world is now definitely warmer than ever before in our (modestly recorded) written history, but it was very much warmer during all sorts of earlier epochs. It was also very cold – at least it was in Europe and North America – three centuries ago, and was that change entirely natural? Hurricanes have become increasingly damaging, with more dollars involved each time a 'big one' strikes, but are we living in more damage-prone circumstances, perhaps pushed there by excessive population and excessive need? And are there not more items to be damaged now, in tune with greater consumption, and also with better inventories to take note of all the loss?

This book, therefore, still investigates the seasons. Our planet revolves, orbits, oscillates, experiences days and nights, summers and winters, heat, cold, drought, storm, gale, calm – all this has always been, and always will be. The major difference since that earlier publication is that we inhabitants have come to realise it is a single world on which we live, the 6,000 million of us who are its humanity. We exist within its atmosphere, a narrow layering which gives us life. What happens to that air, and whether it blows hot, cold, wet, dry, strongly, differently, terrifyingly or even normally, is therefore the basic matter of this book. As to the part we may have played in altering what affronts us, either now or in the future, that is this book's exploratory side. Are we, in truth, to blame? If so, will the amendments being proposed, like cutting emissions by 10%, and being a touch less profligate, make any proper sense?

The Seasons was fun to write back in the late 1960s. Writing about its subject matter yet again, and taking note of all the changes, whether intriguing or frightening, alarming or even calming, has been a great deal more invigorating now that a whole new age has dawned.

A piece of pleading is now necessary – this book is not entirely metric. It is also inconsistent. Children are taught about kilometres, but tend to say that somewhere is 'half a mile away'. Schools teach them Centigrade, but the pupils are happy to stay at home when their thermometers rise 'over 100', giving proper excuse to do so. British signposts use miles and yards, with all limits to speed in miles per hour. The metric system may make more sense, certainly for science, but its words are without the ancient charm of inches, feet and miles.

Inconsistency arises because I quote scientists precisely, with their kilometres and millimetres, or even their kms., cms. and mms. I quote their forefathers no less obediently when they write of degrees Fahrenheit, and would be no less subservient should they mention bushels, furlongs, rods or hundredweight. The aviation industry is undoubtedly modern but universally uses feet for height even when cruising at 800 kilometres in each hour and due to arrive at 1730. In short, the world is inconsistent and I have mirrored that inconsistency. Save for some quotations my heat is in degrees Centigrade, and my distances are in miles. I just prefer days with half an inch of rain to those with 13 millimetres, but I am happy to accept that the sun's surface temperature is 6,000 degrees C. rather than 10,800 degrees F.

In any case it is my book.

A.S.

Chapter 1

PLANET

Planetary fundamentals – aphelion, perihelion – axis – satellites – equinoxes and solstices – new years' days – Easter and other festivals

All our living takes place upon a single ball 8,000 miles in diameter slowly spinning within the emptiness of space. This fact is particularly difficult to accept on some calm evening, when we are relaxing by the rock-like stability of everything around us, and are compelled to remember – as if with a different portion of our brains – that the solid ground and all our immediate environment is simultaneously gyrating and orbiting, quite unsupported save by gravitational forces from elsewhere. If we look at the moon for confirmation of our status, with it too equally positioned by speed and gravity, we can still half-consider that our status on Earth is somehow different.

Perhaps modern individuals are worse at comprehending the planetary situation than our forefathers. Many of them worshipped the sun, and spent more time gazing at the heavens, for inspiration and for guidance, than the modern city-dweller encased within some sprawling and brightly lit conurbation. Take the matter of eclipses, an unpredictable astonishment to most people nowadays, but even early Egyptians and Mesopotamians were attempting to forecast them. There are both lunar eclipses when Earth lies directly between the sun and the moon, and solar eclipses when the moon is between the sun and Earth. The number per year is not constant, with the maximum being seven (mainly solar) and the minimum two (both solar). In every century there are about 66 solar eclipses, their tracks scattered almost randomly across Earth, with a slight bias towards latitudes a few degrees north of the equator. In August 1999 a line from Cornwall to the Bay of Bengal was favoured, and any location will be so blessed, on average, about once in every 360 years. It is easy to wonder how many of the hundreds of thousands who poured into England's south-western county, although enjoying the unfairly cloudy phenomenon, truly grasped the astronomical arithmetic underlying the event. Life on a planet is awkward to absorb.

It was certainly enlightening, as well as exciting, when those astronauts of the 1960s and 1970s returned from their trips to the moon with photographs of our own heavenly body, our single home, our colourful and revolving orb lying within the vastness of space. Those space travellers may have been test-pilots, apparently devoid of ordinary fear and calling Houston unemotionally when a 'problem' was on board, but even they were deeply moved by the sight of Earth. 'That beautiful warm, living object looked so fragile, so delicate, that if you touched it with a finger it would surely fall apart,' wrote James Irwin after his voyage with Apollo 15.

'The entire Earth is but a point, and the place of our own habitation but a minute portion of it,' wrote Marcus Aurelius, most comprehensively, over 1,800 years before the astronauts. It was as if he had seen the pictures which they had collected. These were certainly humiliating, concerning our small-ness, but also awe-inspiring, concerning the vastness of the universe in which we happen to reside. Their blueness was particularly wonderful, making our planet look so welcoming and so excitingly different from every other sight in space. As for the moon, said one astronaut after returning to his warm blue home, it is 'cold and lonely, except when it's hot and lonely'.

Photographs taken by these lunar travellers did inform us, most directly, that we are living on a planet. However calm the evening of our enjoy-ment, however rock-like our surroundings, and however casually some smoke may be curling upwards in our vicinity, thus indicating not a breath of wind, we are also:

gyrating (somersaulting) through 360 degrees once every 24 hours
spinning around our central axis at some 700 miles an hour (or faster
    if we are nearer the equator)
orbiting around the sun at 18.5 miles a second (thus covering 580
    million miles in each year)
voyaging, together with all of our solar system, towards the star Vega
    at 12 miles a second, and
circling within our galaxy around its centre at 170 miles a second.

It is more relaxing for our peace of mind either to forget all this or to believe, as did some of the Ancients, that our secure world of mountains and valleys, of forests and cities, of vertical smoke and horizontal lakes, is actually residing on the back of a turtle (or an elephant, said some Indians, or a frog, said Buddhist lamas).

In fact there is more confusion, for our mental stability, than is caused by all the orbitings. Our Earth is not revolving at right angles to the sun, as if it were a top humming in upright fashion. Instead it is canted over from the plane of its orbit at 66.5 degrees. Consequently our terrestrial north pole is nearer the sun for half the year, and our south pole is nearer to it during each year's other half. For this reason we experience summers and winters, plus all their variations of cold and heat, rain and drought, storm and calm, and either increasing or shortening day length, every aspect of which alters on a yearly basis. (At least our moon is more straightforward. Its axis is almost perpendicular to its path around the sun.)

Worse still, for simplicity, our orbit around our parent star is elliptical, rather than neatly circular. Planet Earth is 3 million miles nearer its sun at perihelion (2 or 3 January) than at aphelion (1 or 2 July). Earth as a whole is therefore receiving differing quantities of sunshine throughout each annual revolution, with inhabitants of the northern hemisphere greatly benefiting from the nature of this ellipse. If the situation was reversed, and the northern hemisphere's winter occurred at the time of greatest distance, all northern winters would be much colder. A 1% increase in light and heat reaches Earth when the sun is nearest, despite the sun then being larger – from our viewpoint – by only 1 minute, 8 seconds of arc. As a further point, our planet is travelling one mile per second faster around the sun at its closest approach. Its orbital speed is then 19 miles a second rather than a mere 18 miles per second when furthest away.

Neither this speed nor the increase in heat is detectable by ordinary citizens, but of course we all notice, due to the Earth's rotation on its axis, that night always follows day. With our axis canted at those 66.5 degrees this causes each day's length to change, the nights being longest in the northern hemisphere on 21 December and shortest on 21 June. These dates of summer and winter solstice (standing sun) therefore do not match precisely the times of nearest and furthest approach to the sun. There is no reason why they should, as the two occurrences are quite irrelevant to each other.

Only those places on Earth either north of latitude 66.5 degrees North or south of 66.5 degrees South ever experience days or nights longer than 24 hours. Similarly, and also due to the Earth's canting, only the area between 23.5 North and 23.5 South (the tropics) ever experiences the sun vertically overhead. Elsewhere, even at midday during the peak of summer, the sun is at a lower angle than 90 degrees. Within the arctic circles, despite the days of summertime being weeks or even months long, the sun never rises high in the sky. It always circles close to the horizon.

As most of Earth's inhabitants live within the 43 degrees of latitude in each hemisphere lying between the tropics and the arctic circles these two regions tend to be viewed as normal. In fact, as they embrace only 86 out of the total complement of 180 latitude degrees lying between the south and north poles, these temperate areas – which are neither tropical nor arctic – are exceptional. They never have the sun directly overhead nor do they ever experience sunshine or darkness for periods longer than 24 hours.

The launch of Sputnik 1, and the subsequent orbital path of this 170 lb object around our planet, forced many of Earth's citizens to wonder how such a man-made contrivance could possibly stay in position. This bewilderment existed despite our own natural satellite, the moon, being out there as solid precedent. Why, if the bleeping man-made thing was truthfully travelling at a reported 18,000 miles an hour, did it not fly off into space or come plummeting back to Earth? What force was propelling it, and then propelling all of its successors – even if we already know that nothing propels the moon, our natural companion?

Many of Earth's inhabitants, however aware of their own orbital track around the sun, found it difficult to appreciate that a combination of speed and gravity, and no propulsion system, keeps all satellites in place, whether artificial like the sputniks or natural like the moon, such things being in continuous free fall. The only difference with the moon's orbit around Earth lies in the moon's distance from its parent planet. It travels at a more leisurely 2,000 miles an hour or so to maintain its orbiting position while circling the Earth.

For perplexed citizens, struggling with sputniks – and why, having passed overhead, they do not pass equally overhead in similar style one orbit later but actually pass some 22 degrees to the west – there then came geo-stationary satellites. These also orbit the Earth but are able to maintain their place in the heavens, as if fixed over one Earth spot (more or less). At 23,500 miles above our planet their speed around us precisely matches the Earth's speed of rotation. Therefore they seem poised over one location, despite travelling in their orbits at some 7,000 miles an hour.

Human beings, notably the scientific ones, like to measure things and detect order where only chaos seems to reign, but all of us – it is easy to suspect – would prefer the Earth to orbit the sun in a set number of days, say 365, instead of 365 and a particularly awkward fraction. Most unhelpfully, for all humans wishing that everything was neat and tidy, or neater and tidier than it is, the precise time of each Earth orbit is 365 days, 5 hours, 48 minutes and 46 seconds. In decimal terms each year is therefore 365.242199 days long, a not terribly convenient number. For simplicity we would also like our moon to be full, say 12 times a year, instead of revolving on its axis once every 27 and a bit days. We would certainly like each lunar month, from new moon to new moon, to last for a more convenient time than 29 and a bit days – or, decimally, 29.5305882.

We would also appreciate a moon which either showed us all of its surface, or a neat quantity like 50%, instead of the 58% plus a bit which is the most we can ever see of it from anywhere on planet Earth. The moon's oscillations (or librations) from side to side, as well as backwards and forwards, are responsible for that odd total of 41% and a bit of moon surface which is permanently averted from our gaze. For additional simplicity we would certainly prefer a spherical Earth instead of an oblate spheroid, one bulging at the equator. And we would definitely like our four official and astronomical seasons to be of uniform duration, set as they are between the equinoxes (when day length equals night length) and the solstices (when Earth's axis is pointing either nearest or furthest from the sun).

By astronomical yardsticks the four seasonal periods are:

Spring (from March equinox to June solstice) – 92 days, 20 hrs
Summer (June solstice to September equinox) – 93 days, 15 hrs
Autumn (September equinox to December solstice) – 89 days, 19 hrs
Winter (December solstice to March equinox) – 89 days, 19 hrs

(At once an apologetic digression is necessary. This book, although written by a northerner, is aware that summertime in the north is wintertime down south. It knows that Christmas-time on Bondi beach is hot, with snow in short supply, that anti-cyclones revolve clockwise in the north and anti-clockwise in the south, that hurricanes twist anticlockwise when heading north-west in the northern hemisphere, and willy-willies in the south are like anti-cyclones in the north, but sentences can become difficult if always making such contradictions. *New Scientist* once chastised itself for wanting to identify a 'summer' issue packed with talk about sun-blockers, holidays and heat. 'We've discovered that hemispheric terms make our Australian readers very angry, to say nothing of our loyal following in Tierra del Fuego . . . So welcome to the late-July special issue.' This book is similar, in having no wish to be accused of hemispherism, as this latest concern is called, but is surely guilty on occasion, the words summer and winter being so engrained, as with spring and autumn. Therefore apologies as and when these are necessary.)

Because Earth is closest to the sun at the beginning of January our planet, as already stated, is then moving fastest. That is the reason why the time from each northern autumn equinox to the northern spring equinox is a few days shorter than the time from each northern spring equinox to its counterpart in each northern autumn. It is also the reason why northerners receive a little more sunlight every year than those living south of the equator. At a latitude of 50 degrees North, for example, the extra ration of daylight every year totals about six hours.

If we were living on a simple spherical body, which travelled in a circular orbit around the sun, every particular latitude on Earth would receive the same amount of sunlight, but Earth's odd shape and elliptical orbit cancels such straightforwardness. The existence of our atmosphere also alters the amount of light reaching Earth's surface. Refraction of sunlight permits us to see the sun after it has actually sunk below the horizon. On the equator this provides about four more minutes of sunshine every day. Further north, or south, this extra quantity increases, being at its maximum where the path of the sun makes its shallowest angle with the horizon.

Ordinary and climate-conscious people will consider the astronomical definitions of seasons to be faulty, with spring firmly concluded in northern temperate regions long before the summer solstice of 21 June, but no two people will ever agree when spring either begins or ends, whether in a normal year or – as all years are odd to some degree – any particular year. Besides, it makes a great difference where these people live. In many places cuckoos will have arrived from their winter quarters, some birds may already have hatched a clutch, and daffodils will be but a fading memory long before the rivers of north-eastern Siberia begin to flow again after their lengthy winter sleep. Some of these rivers, such as the Kolyma, do not start to move until early June (when dead people can, at last, be buried). Hudson Bay, Canada, tends not to lose its covering of ice until mid-July.

Human beings cannot, whether as communities or nations, even agree when the year itself officially begins, with many preferring the equinox of 21 March to 1 January. Moslems start their 'era' from 15 July AD 622 when Mohammed fled from Mecca to Medina. Their new year's day is therefore the first day of the month of Muharram, and there are 12 lunar months to each year, alternately of 29 and 30 days. Each Moslem year lasts for 354 days and 8 hours, which means that 32 and a half conventional (astronomical) years have to pass before Moslems are once again celebrating new year at the same time (by other people's yardstick) and the fasting month of Ramadan is being celebrated at the same time (in other people's years). The requirement to fast between sunrise and sunset during Ramadan is a far greater penance when days are both long and hot, as in midsummer. Jews consider new year's day to be the first day of the month Tishri, which falls sometime between 5 September and 7 October. Most Europeans think 1 January to be entirely suitable, and all the countries peopled from Europe think in similar fashion. Much of this European world, particularly now, is of the opinion that the arrival of 2000 was exciting, being the start of a new millennium, but the same year is also dated 1994 (in the Ethiopic system), 5760 (Hebrew), 1378 (Persian), 1716 (Coptic), 2544 (Buddhist) and 1420 (Islamic). It would be 2753 rather than 2000 if the old Roman system had been perpetuated, and 2749 if the original Babylonian calendar had been maintained.

The notion of a millennium formerly possessed more significance than the arithmetic arrival of another thousand-year period. Some people thought, particularly during AD 999 as 1000 approached, that the change would presage the end of the world. Viking invasions of the period were heralded the Antichrist, an awesome token of impending apocalypse, with bloodcurdling visitations from overseas a powerful portent of doom to come. On the night of 31 December 999 many individuals (certainly in Britain) gathered by crosses or within churches, expecting – and dreading – some form of culmination. The Book of Revelations firmly stated: 'And when the thousand years are expired, Satan shall be loosed out of his prison.' When nothing happened, and 1 January 1000 dawned in standard fashion, there was not only rejoicing but a reinvigorated faith.

Coincidentally there was chiliasm, after the Greek for thousand – as with kilo, etc. Its numerous followers believed that 1,000 years of the Christian era would herald a time when all human flaws would vanish and only happiness would prevail, this dramatic change following a period of tremendous calamity, as with those Viking incursions. Today's population of the third Christian millennium is only too aware of considerable and further calamity throughout the past 1,000 years, culminating in the 20th century's horrendous wars, the most destructive of all time, but there was little expectation when the new millennium arrived that human flaws were about to alter and be replaced by universal happiness after chiliasm had had, so to speak, another bite at the apple.

Precisely nothing of astronomical interest occurs on 1 January, whether or not the date introduces a new year, another century or even a millennium. This prime date is merely ten (or sometimes nine) days after the winter solstice and, coincidentally, a day or two before the closest approach of perihelion. As for the birth of Christ, the actual trigger for new millennia and the basis for today's Christian-based calendar, that date was not calculated until over five centuries after the event. Dionysius Exiguus, a Scythian living in the 6th century, decided in about 525 that Christ had been born on 25 December in the year 753 of the Roman dating system. This scholar also determined that Christ's arrival formed a better start point for the yearly calendar than Rome's reputed foundation almost eight centuries earlier. Several more centuries were to pass before most Christian countries adopted his suggestion, the dating of which led to so much excitement on and after 31 December 1999. This third millennium therefore began in the days of nuclear power, mobile phones and personal computers rather than, if the Roman system had been maintained, the year 1257 when Henry III was on the English throne and the European crusaders had finally lost Jerusalem.

According to Dionysius there was no year 0, no 'year dot' if that is what this expression means. There was only a smooth transition between 1 BC and AD 1, the break-point which he initiated. Unfortunately few people accept that the man who dared to date Christ's birth did his sums right. Many other dates have since been suggested for that event, both before and after AD 1. Some say that 7 January 1998 should have been the start of the new millennium – if the system adapted by Pope Gregory (about which more later) had been still in force. Others, bearing in mind that lack of a year AD 0, feel the world should have waited before celebrating its third millennium until the end of 2000, until 2,000 years had actually occurred. In truth the festivities held during 31 December 1999, and all the (modest) confusion on 1 January 2000, marked nothing whatsoever. No exceptional astronomical event happened on that date – and how nice it would have been had an eclipse occurred or a comet had arrived. Nor had anything outstanding taken place on Earth exactly 2,000 years beforehand. As a final point, for those who believe that Christ was born on 25 December (1 BC, AD 1 or some other year), they have to accept that the Bethlehem happening did not occur exactly 2,000 years before any January 1st.

As for new year's day in general the ancient Greeks thought that autumn, and the beginning of September, formed a more propitious time to start each year, while similarly ancient Romans preferred 1 March, along with the Celts and Druids. It was Julius Caesar and his astronomers who threw their combined weight behind 1 January, this being the first day of the first month after the northern midwinter of shortest days and longest nights. The date was also conveniently in tune with the popular deity Mithras, whose followers celebrated his birthday and that of the unconquered sun shortly after this solar re-birth had been confirmed by slightly lengthening days. That same important time of year therefore achieved greater renown when Christians chose not to remove an established and popular festival but adopted the annual beginning of lessening darkness as a suitable occasion for celebrating Christ's birth. (There is nothing in any of the Gospels indicating the season of that all-important occasion in Bethlehem, save that taxes – the end of the financial year? – then had to be paid. Was it hot or cold in the manger, and was Joseph the only one to fail to find a room?)

With Christ's birth being the initiation of so much it did seem a suitable season to start each year, even though most of a northern winter's chill occurs after that date. Spring, marking the end of winter, might have been considered more appropriate, with plant-growth, egg-laying, and increased warmth such positive harbingers of a new year, but the passing of the winter solstice, and the conclusion of celebrations marking that dark time, are now firmly engrained, in general, as the beginning of another year.

China, a country traditionally rich in astronomers and equally beset by wilful emperors, has been both advanced and also retrograde simultaneously. Its astronomers considered that each new year began when the sun entered Capricorn, namely the winter solstice. The administrative Chinese recognised 12 months – Tzu, Chou, Yin, Mao, Chen, Su, Wu, Wei, Shen, Yu, Hsu, and Hai – but it was up to the emperor to dictate, for all manner of civic purposes, when each new year did actually begin. Therefore, over the centuries and dynasties, various months have had that honour, but gradually the second month became more favoured, and there it has remained.

There is, to make this point once more, no reason why any of the best known astronomical events – Earth's orbit round the sun, the moon's orbit around Earth, the other planetary orbits, and Earth's own revolutions – should bear any relation to each other. Moreover they are all altering in their periodicities, and always have done. Earth, for example, is slowing down in its gyrations. Recent work has indicated a reduction of 2 milliseconds every century. Such a modest alteration only becomes more interesting when multiplied by the great quantities of time our planet has experienced. Such a lessening in speed, assuming consistent deceleration, equals 2 seconds every 100,000 years, or 20 seconds every million years, or 5 hours, 33 minutes and 20 seconds every 1,000 million years. Earth itself is thought to have originated some 4,500 million years ago. As for the origin of life, and the point at which it began during our planet's ancient story, the terrestrial spinning was undoubtedly faster than it is today, with days and nights beginning and ending much more abruptly. The moon was also different. There was assuredly a time when it was spinning faster, and when all of its surface could be seen from Earth – by whatever life forms then existed.

One Babylonian astrologer, working in 136 BC, unwittingly provided confirmation for today's scientists of Earth's slackening in its revolutions. At 8.45 in the morning of 15 April he observed an eclipse, and described it in cuneiform on clay: 'At 24 degrees after sunrise – a solar eclipse . . . The sun threw off the shadow from south-west to north-east.' Modern astronomers, ably assisted by their computers, have determined that the eclipse in question could not have been observed from Babylon. Its zone of totality must have passed through a region 48.8 degrees west of the diligent astrologer. The only explanation (short of error, deceit, fraud, etc.) is that the Earth has slowed down, with the day lengthening by a few milliseconds each century.

When scientific man arrived, the successor to that early Babylonian, and measurements were made of everything measurable, there was expectation – and hope – that some sort of astronomical harmony might be detected, some relationship, some sense in all the various timings. Instead, as was gradually realised, the Earth, the moon, the sun, the planets and so forth are each at a particular stage in their existences, being slightly slower or faster or nearer or further from each other than they were a few millennia ago, or will be a few millennia hence. There is no reason why round or satisfactory numbers should be involved in any of their various and unrelated properties, any more than measurements of islands, trees, animal numbers, continents, oceans and so forth provide convenient quantities.

Nevertheless, as soon as humankind became more sophisticated than simple hunter-gatherers, it was important for those in charge to keep track of the passing days. Festivals had to be correctly celebrated at the appointed times. Moon-based events, the precursors to Easter and Ramadan, had to be defined. Crops had to be planted during the most suitable seasons. Migrations and spawnings had to be anticipated. Farmers, priests, fishermen, hunters and travellers all needed, for their independent reasons, to know the time of year, and how much of each year had already progressed. So, in the main, do all the animals – when to initiate courtship, when to mate, to migrate, to fashion nests and dens. It is somewhat humiliating for humans that creatures such as birds and mammals, and even substantially lower down the animal tree, can be remarkably accurate, being more attuned to changing day-length, in general, than whether climatic conditions are being unseasonably fair or foul.

By way of parenthesis, and to illustrate how confusing humans can be when fixing their festivals in relation to astronomical events, the Church of England's *Book of Common Prayer* describes how to identify Easter Day – in five paragraphs, preceded by a Table. This 'contains so much of the Calendar as is necessary for the determination of *Easter*; to find which, look for the Golden Number of the Year in the First Column of the Table, against which stands the Day of the Paschal Full Moon; then look in the Third Column for the Sunday Letter, next after the Day of the Full Moon, and the Day of the Month standing against that Sunday Letter is *Easter-Day*. If the Full Moon happens upon a Sunday, then (according to the first Rule) the next Sunday after is *Easter-Day*.' And that is the conclusion of the first paragraph.

Once again, as with Mithras' Day becoming Christmas Day, the early Christians took advantage of precedent. Anglo-Saxons used to celebrate the Teutonic goddess of dawn, Eostre, at the springtime equinox, and her name was therefore conveniently transposed to the new festival. Other areas used the root word 'pasch', as in the Italian *pasquale* and the Russian *páskha*, after the Hebrew for passover. In less daunting language than is used in the prayer book another definition of Easter states that it occurs on the first Sunday after the 14th day of the paschal moon, this moon being the first whose 14th day comes on or after the spring equinox (of 21 March). In consequence Easter Day cannot be earlier than 22 March or later than 25 April, its timing having been fixed by the council of Nicaea in AD 325 – about which more in the next chapter.

The calendar is not crucial in talk about seasons and annual events, but time and its various subdivisions can conveniently be placed in a section of its own.

Chapter 2

YEARS AND DAYS

Apologia – Calendar and time – sundials – clock time – 24 hours – 60 minutes, 60 seconds – Universal time and Greenwich time – months – Julian amendments – Gregorian adjustments – leap days – weeks – Nicaean unanimity – menstruation

Everything written about the planetary basics in the previous chapter would be equally true whether or not there were human beings to take note of them. There would be years and there would be seasons in each year. There would, in consequence, be wetter times and drier times, and colder and warmer, and brighter and darker. Human involvement is crucial to this book but human effects upon the climate would be just as relevant even if we had not become literate. It so happens we are creatures capable of science and measurement, and can therefore assess our own mismanagement, but this ability is independent of the mismanagement itself. Some blundering, fire-creating, weapon-wielding, tree-destroying and Earth-provoking ape – not a bad, if brief, description of the species *Homo sapiens* – could be equally at fault even without the power to analyse the error of its ways.

Or without the wish to create and make use of a calendar, subdivided into fragments known as time. Animals and plants do not have such almanacs and yet know when best to reproduce. They too can be destructive, perhaps multiplying excessively and laying waste certain bits of land, but they manage their extremes without chronicles to help with definition. It is therefore unnecessary to examine sundials, clocks, Roman systems, Babylonian history or papal amendments if discussing climate and its quantity of change. The rain would still fall, whether it was doing so at 4.30, on Thursday, or in April. Swallows and other migrants would still arrive and then depart, even if their actions were never set against a man-made diary recording actual dates.

For all those who accept this viewpoint, and who wish to pursue the major subject matter, it would be better to duck this chapter altogether and proceed at once to its successor. For all those others who wonder how it happened that weeks, minutes and seconds, along with months, leap years and other peculiarities have come into being, the following pages may be of interest. Humans are an odd lot, and the system for measuring time that they created for themselves is one of the odder arrangements they have acquired, presumably – but this is debatable – to make life simpler for themselves.

Early history The basic wish for an obliging orbital arrangement existed because Earth's gyrations both initiated and defined our concept of time, with night and day being such blatant happenings, and with each year following its predecessor with similar regularity. Early Egyptians are credited with inventing the sundial (in 1300 BC) and then with subdividing the daytime period into 12. These 12 original parts were faulty as measurers of time because early Egyptian sundial hours varied with the length of day, and therefore as each year progressed. The sundials themselves were gradually improved, most notably when the gnomon (or projection casting its shadow on the sundial's face) was tilted at 23.5 degrees, thereby taking account of Earth's axis. In fact even good and modern sundials only measure what is known as apparent solar time, this differing slightly from clock time.

Sundials are the reason why the hands of all clocks and watches rotate, well, clockwise. With the sun rising in the east, setting in the west, and casting its shadow from the south, this shadow moves in a clockwise direction. Had sundials been invented in similar latitudes in the southern hemisphere, where the sun is always to the north, clockwise would be in the other direction. The time for noon – 12 o'clock – is at the top of a watch/clock because the sundial's shadow is then pointing due north (in the northern hemisphere). Therefore, when clocks were invented, it was logical to maintain the existing custom, for time to travel clockwise and for noon to be on top. (Of course, as pedants like to tell us, a clock's hands go anticlockwise from the clock's point of view!)

Subsequent horologists following those early Egyptians had to find a means for measuring the passage of time when sundials were ineffective, notably at night or when sunshine was obscured by cloud. Therefore sand glasses, water clocks and graduated candles all came into use, thus shifting the concept of time away from a dependence upon day-length. The first mechanical clocks were fashioned in 13th-century Europe, with their inevitable inaccuracies frequently readjusted by reference to the sun, still the dominant reference point. These earliest clocks did not have a minute

hand, but did form the major timekeeping advance until 1656 when the
first pendulum clock was built – by Christian Huygens in Holland. He is
also credited with developing the spiral balance-spring, ancestor of the
modern hairspring.

Mechanical time was not thought to be the real time, hence the
custom for such times to be given 'of the clock' which was then shortened
to 'o'clock'. A further linguistic curiosity lies with the term Greenwich
Mean Time. Clock time, as with GMT, is based on an average length of
day. Yet another strangeness, probably irrelevant to history, is the
custom for people to ask strangers for 'the right time'. What other time
is there?

Perhaps they should instead be asking for 'local time'. The modern
ability to telephone Australia with as much effort as telephoning the house
next door demands that all of us first calculate whether it is a convenient
hour for disturbing distant acquaintances. Are they ahead of, or behind,
our local time? Only a century ago (9 February 1899) *Nature* was
applauding that a series of tables had been produced showing 'the
differences between Greenwich mean time (established as the first global
time scale in 1884) and the civil times used in various parts of the world'.
In an imperial manner, scoffing at conceptions of time elsewhere, *Nature*
added that 'Chinese at most places use an approximate apparent solar
time, obtained from sun-dials. At Tientsin the civil time is determined by
the municipal chronometer, which, however, has sometimes been known·
to have an error of three minutes. The Persians keep sun time, watches
being set at sunset. In Teheran there is a midday gun fired by the time
shown on a sundial. But a few minutes makes no difference in Persia; the
railway trains start when full or when required . . .'.

Somewhere in the story of time's measurement there arose the notion,
no doubt encouraged by each equinox when days and nights are of equal
length, that a further 12 periods constituted night. The numbers of hours
from dawn to dawn thenceforth totalled 24. The advent of am and pm
saved us from ever saying that, for example, the time is 23 minutes past 17
o'clock, but airlines and the like have had to revert to a 24-hour system for
greater overall convenience.

Babylonians are generally believed to have made the next move, by subdividing each twenty-fourth of a day into 60, and then subdividing these
smaller units by a further 60. They therefore created our modern, and not
wildly convenient, system, with each terrestrial revolution lasting 24 hours,
or 1,440 minutes, or 86,400 seconds. Modern metric man now divides
these smallest units into milliseconds, into microseconds and even nano-

seconds, with all of them based upon a system that is totally non-metric. Ten hours per day, with 100 minutes per hour and 100 seconds per minute would have been preferable, each of our terrestrial revolutions then lasting for a neat – and certainly neater – 100,000 seconds, a total particularly satisfactory for communities whose numerals are based on 10. There would then be 100 million milliseconds in a day instead of 86,400,000, today's peculiar and unhelpful quantity.

The second as a unit of time is now too embedded in our lives, our technologies and our clocks and watches to be cancelled in similar style to all the early yardsticks of rod, pole, perch, furlong, etc. being bluntly displaced by the metre, but scientists have cancelled the second's original interpretation. The first clock in history to be regulated by the spin of a molecule, instead of the sun or stars, started 'ticking' in February 1949. It was unveiled at the US National Bureau of Standards and was controlled 'by the period of vibration of the nitrogen atom in the ammonia molecule'. The old Babylonian second, defined as one 86,400th part of a day, was then formally redefined in 1967 as the time taken for a caesium atom to undergo 9,192,631,770 magnetic transitions. That statement may be meaningless for most of us, but modern science needs to measure seconds so accurately that the former system became invalid, not least because Earth's orbit around the sun is – very slightly – altering from year to year.

Coordinated Universal Time (UTC) was adopted as the official time for the world in January 1972. UTC is a compromise between atomic clock time and Earth's rotation time. Consequently, as Earth's time alters, leap-seconds have had to be introduced, and 22 such seconds have been added between 1 January 1972 and 1 January 1999, each change having taken place at the end of either June or December. If there were not these minuscule adjustments the sun would be overhead at midnight – rather than noon – in about 50,000 years' time. The current atomic clocks are extraordinarily accurate, but there are plans to make even better kinds, notably by a system known as ion trapping. Suffice to say that a clock based upon this technology, once it has been perfected, will lose less than 0.0000000000001 seconds per day – or no more than one second in the lifetime of the universe.

The 21st century's concept of time is bewilderingly different from any Babylonian's, but even the start of the 20th century can seem archaic from our current viewpoint. *Nature* regularly includes fragments from its columns of 50 and 100 years ago. On 9 February 1899 it commented on 'a case concerning the lighting of bicycle lamps'. An act of 1888 had stipulated that all such lamps should be shining one hour after sunset and until one hour before sunrise, with Greenwich fixing the hour of these events. A man in Bristol, which is situated over 2 degrees 30 minutes west of Greenwich, was found guilty of being unlit according to Greenwich time but not according to the later Bristol sunset time. *Nature* noted that his conviction was quashed, with local time thenceforth being the time in question, the right time, the real time, the time for lighting lamps.

While searching for a larger unit than a single day, however subdivided, the early astronomers, principally of China and Egypt, noted that the moon took slightly less than 30 days between one new moon and the next, there being some 12 new moons a year. The moon's cycle therefore provided that larger unit, most conveniently, with the result that months – along with days – became fixed as the solid basis of both our time and calendar. The early Chinese, wishing to make even greater sense of lunar cycles, also decided that a thousand new moons equalled one human lifespan. These people were therefore more generous, or optimistic, than the Biblical psalmist who sang of 'three score years and ten'. The Chinese allocation gives each of us nearer four-score years than Psalm 90's ration. Even if the Chinese were thinking of lunar revolutions rather than new-moon to new-moon phases the allowance is still more generous.

Efforts were inevitably made in the early days to match moon cycles with year cycles. Early Romans, for example, considered that each year contained ten new moons. Their list of months were chosen as:

Martius, after Mars, god of war,
Aprilus, after the Latin for opening, with leaves burgeoning,
Maius, named, according to Brewer, not from Maia, mother of
   Mercury, as is frequently alleged, but from Latin for growth,
Junius, not named after Juno, Jupiter's wife, but its origin is hazy,
Quintilis, fifth month,
Sextilis, sixth month, and September, October, November and
December, being the seventh, eighth, ninth, and tenth months.

When this array proved to be inadequate (by a long way, as it totalled only 295 days in one year) two more months were officially added. January was attached at the beginning of each year and February at the end. This first month was named after Janus, the two-faced god who was, presumably, looking back at the old year and forward to the new. February was named after *februa*, the word for expiation. This early Roman atonement for past error therefore mirrored our current custom of new year resolutions, the

Roman arrangement being regret for things already done and the modern custom a promise, however short-lived, of improvement in the future.

The next major Roman alteration, established in 452 BC, was the transposition of February from its important position as conclusion of each year to a less prestigious location as second month. It therefore followed Janus's month, but there was still the old determination to fit a precise number of lunar cycles into each annual cycle, with the 12 named months from January to December possessing 29 and 30 days alternately. That total of 354 days was somewhat arbitrarily boosted each year by one extra day, thus making 355, and the consistent shortage (from 365) was made good – or rather better – by the addition of a new, and relatively short, month every other year. Mercedonius, interposed between the 23rd and 24th of February, was either 22 or 23 days long. This new kind of Roman year, although an improvement on its predecessor, was not only much more complex but it still created a wrong total, namely 366.25 days per year. The slight excess of one day each year was sorted out – more or less – every 24 years by the subtraction of Mercedonius.

This revised calendar was still plagued by the old desire to match moon cycles with Earth cycles, to marry months with years. Such persistent wrong-headedness was corrected, four centuries later, by Julius Caesar working together with his Greek astronomer Sosigenes. They jointly decided to tidy up the extremely irregular – and most un-Roman – system. After much deliberation they retained the 12 basic calendar months, now with names and ordering too well entrenched for alteration, but in making all save one of these months last for either 30 or 31 days the new 'Julian' system no longer paid any real heed to the moon's actual phases and cycles. Every month, whether long or short, was longer than any lunar cycle.

In future, as was officially stipulated in that first century BC, there were to be six long months – January, March, May, July, September, and November – which were to be neatly interposed by six shorter months – February, April, June, August, October and December. It must have irritated the Roman mind that each year's ninth month – September – was therefore named for *septem*, Latin for seven, and that October, November and December were equally mistitled, but worse was to come. Had Caesar and Sosigenes wished for slightly different regularity they could have arranged for seven months of 30 days plus five months of 31 days. This would then have totalled a convenient 365 days, but they opted instead for regular alternation, save for the second month. Poor February, already demoted from its important post of atonement which concluded every year, now became the chop-and-change month, usually with 29 days but with 30 days every fourth year. The end result of this Julian amendment, formally initiated in 46 BC, provided an average of 365.25 days a year, this being decidedly more accurate than the earlier 366.25 days, with Earth taking 365 and almost one-quarter days to travel round the sun.

The so-called 'year of confusion', 46 BC, was most aptly described. It not only possessed the extra – and soon to be discarded for ever – Mercedonius, but two additional months which had never been seen before and were never to be used again. This 15-month assortment brought that year's total of days to 445. Following its undoubted confusion almost everyone agreed that earlier error had been decently amended, with the decisions of 45 BC marking the start of a much better arrangement. (Alas, but Julius Caesar did not long enjoy either the glory or the satisfaction of this undoubted benefit. Disgruntled republicans killed him on the Ides of March one year later.)

However, as a mark of gratitude for his calendar endeavours, and conveniently neglecting Sosigenes' contribution, Julius's name was not only given to the new arrangement, the 'Julian' calendar, but it officially replaced that of Quintilis, the former fifth month. In consequence we now have July/Juillet/Juli/Iyul/Yuli/Luglio, etc. It so happened that this seventh month, Caesar's eponym, was one of those to which he had allocated 31 days. Augustus, successor to Julius and first Roman emperor, welcomed the precedence of a month being named after the man in charge. It was therefore decided on his behalf that the former eighth month, next in line, should become Augustus month. As this eighth month possessed only 30 days, and was therefore inferior to July, Julius's inheritance, it had to be promoted to become a 31-day month, this change instantly disrupting Julius Caesar's tidy ordering. In consequence, causing further upset but setting the annual total correct once more, September was perfunctorily demoted from 31 to 30 days.

The well-arranged, distinctly sane, and alternate ordering of long-short-long-short months was therefore upset, causing the modest mayhem of today as we each mouth some jingle like 'Thirty days hath November, April, June and September' (or even count one hand's knuckles by another method) in our wish to remember the date. Julius should undoubtedly be praised for his efforts at tidiness, but Augustus – or his sycophants in the Senate – should be soundly berated for untidying them to create today's unkempt anomaly.

At least Tiberius, emperor after Augustus, did not cause further trouble by donating his name, or having it donated, to September, by then a short month which would presumably have needed face-saving promotion and caused additional imbalance to the Julian regularity. In fact, there were to be later moves on behalf of Nero, Claudius and Germanicus to rename April, May and June respectively, but Tiberius had been more than modest, he had been sensible. 'What will you do when there are 13 Caesars?' he asked, most pragmatically. The subsequent renamings did not survive, possibly out of respect to Tiberius's perspicacity.

During the subsequent centuries various scholars noted that Caesar's system of days per year was proving faulty, with each equinox gradually altering from its proper date. Roger Bacon, polymathic scholar of the 13th century, concerned himself deeply with the Julian calendar, along with practically everything else. This English Franciscan friar vigorously pointed out that Christians were celebrating Easter and other festivals on quite the wrong dates, the calendar having slipped nine days since Caesar's reorganisation almost 14 centuries earlier. Challenging the Catholic Church, on anything, was a form of heresy, and the unfortunate Bacon was officially silenced by imprisonment.

Much later Pope Clement IV heard of the friar's work, and asked to see the calculations. Bacon, having been persecuted even by his own order, was naturally overjoyed. He spent two years putting together a suitably fulsome document, before despatching it to Rome. Alas, the sympathetic pope died in 1268, perhaps not even seeing the manuscript and certainly without opportunity to implement any of its recommendations. In the absence of such a patron, and with many enemies determinedly defending the veracity of the Catholic Church, poor Bacon fell foul of authority once again. He was attacked, reputedly for 'certain novelties', and his sensible calendar thoughts were firmly neglected. He probably died in 1292, but no one took the trouble to record the death date of this difficult and brilliant individual.

Caesar's Greek adviser, although skilled and knowledgeable, had considered the astronomical year to be 11 minutes and 14 seconds longer than it really is. This may seem like a small mistake, particularly for his day and age, but the passage of years steadfastly enlarged that modest error. By the time that Gregory XIII was on the papal throne in the 16th century numerous astronomers and even churchmen were recommending that Catholics should mend their ancient ways. Moreover, during the 16 centuries from Imperial Rome to Renaissance Italy, the business of astronomy had greatly improved, causing the Roman fault to become increasingly conspicuous. Hence the build-up of pressure to improve matters, particularly when so much else was being advanced in that general time of enlightenment.

In 1582 Pope Gregory directed that ten days should be suppressed from the calendar, thus restoring the equinox to 21 March. This subtraction of such a quantity of time had a powerful effect upon many of his contemporaries. They felt cheated out of that amount of life, believing their actual date of death had been somehow preordained. Even today, or particularly today, there could be similar upset, with ordinary citizens affronted by such a liberty being taken with their lives, let alone with their precious and bug-imperilled electronic contrivances.

The fact of a pope's recommending anything in 1582 received short shrift in London, with the Protestant Queen Elizabeth on the throne, with the Catholic Queen of Scots still a thorn in her side, with the Church of England determinedly established, and with any edict from Rome inevitably condemned as some further form of popish plot. (The Roman calendar pronouncement occurred 23 years before the Protestant-born but Catholic sympathiser Guy Fawkes was caught attempting to blow up Parliament.)

Therefore Elizabeth's realm did nothing, despite many adjacent European nations promptly adopting the 'new style' dating system. Travel abroad to the continental mainland then meant more than a change of time, today's occasional adjustment. It caused, for anyone departing on, say, 28 April, the date suddenly to become 8 May. And vice versa on returning home, when 2 June was abruptly transformed into 23 May for everyone taking a ship across the Channel. Current historians hate this misalignment, always having to know whether new- or old-style dating was being observed at the time, or was being observed by subsequent historians when detailing some event from that period.

Scotland, being less upset about Roman pronouncements, switched from old to new in 1600 without any great upheaval, save presumably for Scots or English daring to cross the Tweed from one nation to the other. Even Protestant Germany began to make the change to Gregorian dating in 1700, although not completing the switch among its various regions until 1775 – which must have been splendidly confusing. Great Britain as a whole finally fell into line 170 years after Pope Gregory had made his perfectly rational suggestion to bring the equinox once again to 21 March. The British Parliament passed an Act in 1751, when George II was comfortably on the throne and Rome less feared, ordering that 2 September in the following year should be succeeded by 14 September. By then, with the Julian error still being locally perpetuated, the original ten faulty days had become 11. There were serious riots leading up to the change, provoked by that loss of 264 hours of life, with 3 to 13 September simply vanishing, but the government stayed firm and the 'new' dating system was duly instituted. As *The Ladies Diary: or, Woman's Almanack* then put it: '1752 September hath only XIX Days in this Year' because 'The Account of Time has each year run a-head of Time by the Sun.' Succinctly put, and so it had.

Several legacies of that eventual change are still with us today in Britain. Common fields of the manor used to be opened for grazing on 1 August, Lammas Day, the feast of first fruits. The shift to new style transformed this permitted opening to 12 August, creating the reason why grouse-shooting is now permissible as from that otherwise bizarre date. The formal financial year used to end on 25 March because that, being Lady Day celebrating the Annunciation, used to be a form of new year's day. The Gregorian change, when duly instituted, caused that date to become 5 April, another seemingly muddle-headed choice of day. Russia was even more antagonistic than was Britain to any alteration imposed from Rome, and did not make the necessary amendment until, with revolutionary fervour then in progress for almost everything, the 20th century. Japan accepted the Gregorian system in 1873, but China did not do so until, also with revolutionary enthusiasm, 1949.

Pope Gregory's edict did not merely put matters right; it so rearranged the system that further, and future, amendments would become less necessary. The fundamental problem lay, as earlier astronomers had already learned, in the awkward length of time Earth takes to orbit the sun, namely those 365 days, 5 hours, 48 minutes and 46 seconds. If the orbital period had been longer by 11 minutes and 14 seconds, making each of Earth's circuits around the sun last for 365 days and 6 hours, the task of adjusting days to years would have been relatively simple, with every fourth year being a leap year of 366 days. As, unhelpfully, it does not do so, and as Pope Gregory had wished his improvement to be long-lasting, he had ordained that leap years should be retained but occasionally dropped. Or rather that February's leap *days* should be kept and occasionally abandoned, this being the proper name for 29 February.

It was therefore decided that these extra leap-days should occur in all years divisible by four, save for three-quarters of the century years. In short, 1600 and 2000 would be leap years but not 1700, 1800 or 1900. Such a procedure makes calendar years almost coincide with true solar years, with solar years shorter by a mere 26 seconds – or, more precisely, 25.96768. The year 2000 was therefore the first century year in England since the change to be a leap year. Subsequent astronomers – after Gregory's time – have proposed that the 26-second error would be further reduced if the years 4000, 8000, 12000 and all future multiples of 4,000 are treated as standard rather than leap years.

With Julius Caesar's reorganisation, made during relatively ignorant times, lasting for 1,600 years it is possible that the present system, the rightly named Gregorian calendar, may last even longer. If the post-Gregory recommendations, concerning the years 4000, 8000, etc., are not made there will be modest trouble, with the Gregorian calendar ahead of the true solar year by one whole day when the year 4909 arrives. It is unthinkable that our technologically superior descendants of that time will tolerate, even for a nanosecond, such gross inaccuracy.

Weeks Pope Gregory's changes did nothing to remove months, however wrongly based upon lunar cycles. They also did not interfere with weeks, another human arrangement which synchronises with nothing what-soever. Even Julius Caesar, despite his considerable improvement to the calendar, took no formal note of weeks. They were not officially intro-duced into the Roman calendar until the time of Theodosius in the 4th century AD. Weeks do not relate to, or fit neatly with, moon cycles. Neither do they have any links with Earth's cycles around the sun, whether in normal years (52 weeks and 1 day) or leap years (52 weeks and 2 days). Although they are man-made entities, usefully dividing up the annual cycle, their seven days can never form a whole fraction of each year. That could only have been achieved – with 365 being such an unhelpful number – by creating 73 weeks per year with each week lasting five days. Or, conversely, having five weeks in every year with each one lasting 73 days. No other multiples of seven fit into 365, both numbers being so awkward.

The notion of seven days forming a single unit is thought to have originated in Mesopotamia, that oasis of advancement during many early centuries. King Sargon I of Akkad, near modern Baghdad, who lived from 2335 to 2279 BC, is believed to have decreed that the seven-day week should reign within his empire. He also established the first Semitic dynasty and defeated the Sumerian city states, but that definition of a week lasting seven days has given him a more emphatic inheritance, one which is now over four millennia long.

Why this king chose seven as the all-important number is unknown. Perhaps the seven neighbouring heavenly bodies, namely the five observed planets plus the sun and the moon, were the root cause. Or perhaps seven, even then, had a powerful hold upon humanity. Over the centuries, and in all manner of locations, there have been (with thanks to Anthony Michaelis for his collation in *Interdisciplinary Science Reviews*) the Seven Sisters (of the Pleiades), the Seven Wonders of the Ancient World, the Seven Wise Men (of Greece), the Seven Sages of the (Chinese) Bamboo Grove, the Seven (Japanese) Gods of Luck, the Seventh Heaven (of the Muslims), their Seven Walks around the Kaaba, the Seven-Candled (Jewish) Menorah, and the Seven Hills of Rome. The Christian religion has the Book of Seven Seals, the Seven Champions of Christendom, the Seven Deadly Sins, the Seven Holy Virtues, and the Seven Spirits before the Throne of God.

Perhaps King Sargon was being influenced by astrologers, or by the inherent magic of the number seven, or merely by the thought that seven days formed a convenient interval between one market day and the next. At all events, and thanks to Mesopotamia in general, the notion of seven-day periods was well entrenched in human lives long before the Romans came to power, and certainly before Christianity began. This fact is not particularly clarified in *Chambers' Dictionary* which, in defining hebdomad, states that it is a 'set of seven: a week: in some Gnostic systems, a group of superhuman beings'. The *Oxford Concise Dictionary* defines the same word as 'Week, esp. in reference to Dan. ix. 27' – about which more in a moment.

In the western world these seven days were named after the seven nearest heavenly objects, and we still have, in English and various other languages, Saturn's day, and the sun's day, and the moon's day. The days from Tuesday to Friday were appropriated in Germanic Europe for various Norse deities, with Tyr (brother of Thor) for Tuesday, Woden (the Anglo-Saxon form of Odin) for Wednesday, Thor (son of Odin) for Thursday, and Freya (wife of Odin) for Friday. Tuesday in the old style used to be Mars day, and still is with Latin nations, as with the Italian *martedi* and the French *mardi*. Wednesday was Mercury's day – *mercoledi*, *mercredi*, Thursday was Jupiter's day – *giovedi*, *jeudi*, and Friday was Venus's day – *venerdi*, *vendredi*. Just as Mithras' day became Christ's day, thus causing minimum change and inconvenience with one festival being replaced by another, the various weekdays were incorporated into religious systems rather than overthrown, with the Sabbath (for Christians), the Shabbat (for Jews) and Djum'ah (for Moslems) becoming the holiest days of each 7-day week – although on Sunday, Saturday and Friday respectively.

Weeks are only rarely mentioned in the Old Testament, as in that Chapter 9 of the Book of Daniel: 'Seventy weeks are determined upon thy people and upon thy holy city, to finish the transgression, and to make an end of sins . . . Know therefore and understand that from the going forth of the commandment to restore and to build Jerusalem unto the Messiah

the Prince *shall be* seven weeks, and threescore and two weeks: the street
shall be built again, and the wall, even in troublous times. And after
threescore and two weeks shall Messiah be cut off, but not for himself . . .
And he shall confirm the covenant with many for one week: and in the
midst of the week he shall cause the sacrifice and the oblation to cease . . .'.
In all of this complex statement there is no indication of one day in each
week being a Sabbath. Neither is there indication of the number of days
which made up one week.

In Genesis, most famously, six days were recorded as the time taken to
create 'the heavens and the earth', with God resting on the seventh day,
this causing that day to be 'sanctified', but there is no affirmation that the
week which we all use was created, and made significant, because of those
six days of labour and one of rest. The word 'week' is not mentioned
anywhere in Genesis. It is therefore generally presumed to be no more
than a very ancient wish by humankind to keep track of passing days.

Constantine, more than any other individual, took hold of the old Julian
inheritance and amended it to embrace Christianity. This involved a
consolidation of the seven-day week. It became an integral part of the
calendar, although it had nothing to do with months or years. Sunday also
became the official day of rest, the holy day, the sabbath. Constantine
therefore overturned the traditional concept, favoured by Jews and
Roman pagans, of Saturn's day being most important. He threw his
weight behind the Sun's day, which was already significant for the
Mithraist sun-worshippers. Some of the early Christians had long
preferred Sunday, mainly because Jesus – the light of the world, and
therefore the sun – had risen from the dead on the first day of the Jewish
week, namely Sunday. Constantine was being both pragmatic and
positive, favouring what many already favoured, and unifying the various
Christian factions.

He also clarified the problem of Easter. This moveable feast, as against
the fixed feasts of Christmas and the like, had been celebrated on all
manner of different days. Before his time no one could agree when the
earliest Easter had taken place, the Easter of the crucifixion, the most
critical of Christian dates. It was plainly important that all Christians
should venerate that event simultaneously. At Nicaea, near modern
Istanbul, Constantine took charge in June AD 325 in a most determined
fashion. He collected all the written statements from all the various
factions, each putting forward distinct points of view about Easter and
much else, and then threw the whole lot into a fire. (Many a current

chairman of some squabbling committee would, most assuredly, love to follow suit, and a few may even do so.)

Following this ruthless approach, and when the gathering at Nicaea was being finally concluded, Constantine was able to proclaim: 'By the unanimous judgement of all it has been decided that the most holy festival of Easter should be everywhere celebrated on one and the same day.' Easter's actual timing is not significant, however intriguing, in a book about seasons, but the calendar *is* important. It dictates how we measure the passing days, as well as months and years. (If anyone wishes to read more about such quirks as Easter, or calendars in general, David Ewing Duncan's book *The Calendar*, first published in 1998, is warmly recommended.)

Many a modern commuter, hurrying to town for a five-day week, and then hurrying home for a two-day weekend, might prefer a ten-day week, as the ancient Greeks enjoyed, and then a three/four-day weekend, but the seven-day week is now such a solid inheritance that change is probably impossible. When the 19th century was coming to an end, and calendar thoughts were in the air as they have been recently, Mr A. Hall of Highbury was ingeniously suggesting that, thenceforth, New Year's Day should be called 0 January. (For many individuals, having celebrated excessively on its eve, the day is, in any case, a bit of a blank.) The remaining 364 days should then be divided into 13 months each of 28 days. If his advice had been followed any particular day of the month would always have fallen on the same day of the week. The 9th of any month, for example, would always have been a Tuesday, as would the 16th and the 23rd. He further recommended Christember as a name for the 13th month, a choice possibly not acceptable universally, but there was much to be said for his general idea. No wonder, therefore, it was never adopted. Or maybe no one could tolerate such a degree of change, however sensible. During World War Two some sections of the American army based in England instituted an eight-day week in the run-up to D-Day, partly because there was so much work to be done but also to ease the pressure on local recreation centres by using them on different days of the week, with the eighth day being a rest day generally out of tune with the nation in which they laboured.

We today are living in quite a different age, even to World War Two. We have been putting footprints on the moon and are contemplating trips to Mars; but, when human beings do blast off for a different planet, the event will take place during a certain week and within a particular month. Both these quantities of days were created by forefathers, who were also looking at the heavens for inspiration, and who never for a moment could have imagined that their terrestrial descendants might travel from the Earth which gave them birth. If those ancients had achieved such a mental leap they would surely never have presumed that such technologically advanced, and therefore bewilderingly different individuals, would be using a calendar which they had instituted so many centuries before space voyaging could be contemplated, let alone achieved. Perhaps the first visitors to Mars will depart from their parent planet on Mars day, a Tuesday. It would be happily appropriate, and there is even precedent. Neil Armstrong and Edwin Aldrin, first human visitors to the moon, climbed down a ladder to reach its surface on 21 July 1969. It was indeed a Monday.

As a further connection between the moon and Earth's inhabitants, or rather to half of them, the moon has also given its name to menstruation, to menses and the menopause, but no relationship exists between lunar cycles and those experienced by women. Women living together may have their menstruations synchronised, as if some controlling agency, such as the moon, is at work, but pheromones are to blame, these being secretions produced in association with the menstrual cycle.

The early human longing to find some sort of affinity with the heavenly bodies, and to make sense of all their spatial wanderings, has given us an intriguing legacy of words and timings, with the calendar being the greatest and most curious mish-mash of them all. It possesses 52 and a bit weeks per year, as well as the 12 months which have nothing to do with the moon, varying as they do in length from 28 to 31 days. These subdivisions of each year always add up to 365 days – except, of course, when they add up to 366.

As an aside there are quaint parallels between this calendar inheritance and the former British currency. There may once have been sane reason for its various enumerations, such as pence, shillings, tanners and crowns, so astounding to all foreigners and pleasing to the locals, but modern times overthrew them in 1971, their arithmetic too complex for a world of decimals. It would appear that timing and the calendar, with their equally awkward quantities of 60 minutes in each of the 24 hours per day, with seven days per week, and with a day or two more than 52 weeks per year, leading to 365.25 (almost) days in every revolution round the sun, will be a tougher nut to crack. There was considerable talk, particularly from decimal currency countries, that Britain should change its style of money long before it did so, but no one seems to be suggesting that weeks or months should be replaced, or even modestly amended, for our greater convenience.

We today live in such a different age to those times when weeks and hours were initiated, but whatever we do today, however sophisticated, however brilliant and entirely novel, still takes place on a particular weekday and at so many minutes and seconds past the hour. Those originators would be amazed by current living, but would be flattered that so many millions of us go to work on the moon's day, and rest not only on the sun's day but, if we can, on Saturn's day as well. The Mesopotamians, in particular, would be additionally intrigued that we each carry a clock which tells precisely how many lots of 60 seconds and groups of 60 minutes, and how many of the 24 hours, have passed since the previous day. Our physical bodies are identical, as near as dammit, to those of our industrious ancestors. The several arrangements they initiated concerning the passage of time, per day, per year, are also absolutely our firm inheritance.

Chapter 3

Basic Situation

The sun – millions – solar radiation – ultraviolet – eclipses – solar future – spherical effects – atmosphere – land and water – surface furnishings – lows and highs – winds and calms – micro-climates – tree line – natural variation

Earth's life-giving star is the provider of most of our light and all but a minute fraction of our heat. Some light reaches us from other stars, and some is reflected our way from the moon, but the sun is the primary source of illumination. As for being a heat provider its radiated warmth is not our total source, with a very little – less than one-tenth of a watt per square yard of surface – being produced by the Earth itself. The sun's bulk is massive relative to Earth's, its diameter being 109 times greater and its volume 1.3 million times more bulky. That diameter is therefore almost equal to twice the diameter of the moon's orbit around Earth, and the sun's volume is 337,000 million million cubic miles. This heat provider is relatively lighter than Earth, with each of its cubic miles being about one quarter the weight of each terrestrial cubic mile. Nevertheless its greater mass dictates that gravity at its surface is 28 times more powerful than on Earth, and a man would weigh two tons if he were able to stand there.

The sun's heat is marginally supplemented by the Earth's own warmth. During the 19th century some of the very best of physicists, having measured the temperature of deep mines and of flowing lava, decided that this inherited warmth must be lessening speedily, with the planet swiftly cooling. The fact that terrestrial temperature increases downwards at the rate of about 30 degrees Centigrade for every kilometre of depth (3,280 feet) prompted William Thompson, later Lord Kelvin, to calculate Earth's age as 200 million years. His result, made public in 1862, thus confronted many geologists – Charles Lyell in particular, whose *Principles of Geology* had so impressed Charles Darwin. In 1899 Kelvin chose to reassess the planet's age, and gave it an even shorter life, despite Lyell's opinions then being widely accepted, despite Darwin's *On the Origin of Species* having been published 40 years earlier, and despite much scientific opinion of the time affirming that Earth was a much more ancient place.

There was a further difficulty concerning Genesis and the biblical story of creation. The Bible's opening chapter still had a literal hold upon many minds, and even James Ussher's words continued to carry weight. This 17th-century Irish churchman had decided that the creation of the universe took place on 22 October 4004 BC. By no means was his dating universally believed, even among the devout, but there was widespread concern that science was being a little too cavalier in overthrowing so many biblical statements quite so wholeheartedly.

The Kelvin belief in a cooling planet, warmed by the sun but rapidly losing its original capital of warmth, ran counter to the geological (and palaeontological) notion of a very much older planet. Lyell, having looked at sedimentary rocks in particular and at the Earth in general, postulated that hundreds of millions of years might be more reasonable. The evolutionists were of a similar view, unable to believe that natural selection, after this had been proposed, was able to act as swiftly as Kelvin had suggested. Unfortunately the great scientist – and he was great – did not take radioactivity into account for the very good reason that it had not yet been discovered. That had to wait until Antoine Becquerel investigated uranium, Wilhelm Roentgen discovered X-rays, and Marie Curie, after working on radium, added to their discoveries and coined the name radioactivity, all of this occurring in the 1890s.

Kelvin did not know that heat is generated when certain atoms spontaneously disintegrate, a procedure now tamed with large atoms in nuclear power stations and explosively demonstrated in atomic bombs. Earth will cool, in time – much as the moon has done – but nothing like so hastily as Kelvin had propounded. Current thinking is that about half of Earth's cooling is coming from radioactivity and half from its original heat. The more it cools, as with any cooling body, the longer will be the time between each loss of one degree. Work published in September 1999 stated that Earth's central temperature is about 5,500 degrees C., which is roughly that of the sun's surface. Much internal rock is liquid, with volcanoes affirming this point when pouring forth their lava, but the very centre is solid as the pressure is so great. This internal core extends from Earth's central point to about 55% of Earth's radius, namely 1,800 miles below the surface. The outer core is liquid, and is thought to provide sufficient heat to influence mantle convection and therefore the movement of tectonic plates (about which more in the section on earthquakes).

The sun's brightness is equivalent to 1.5 million candle-power per square inch of its surface. Its internal temperature reaches 15 million degrees but the surface temperature is a mere 6,000. However its corona, the disk of light around the sun, has a temperature of several million degrees, despite being further from the core than the solar surface. Magnetic loops lashing out from that surface are thought to despatch enough energy to the corona to explain its greater temperature. Each loop 'carries about as much energy as a large hydroelectric plant will generate in a million years', stated a scientist attempting to illustrate the scale of this phenomenon.

For those of us not intimately acquainted with millions of anything, and with that number already peppering this chapter, it is possible to lose track of the word's true significance. One million is a very big quantity. Or rather it is a quantity which lies beyond the experience of most human lives, save for financial executives, demographers, exchequer chancellors, and their like. The standard human life span from birth to death is some 27,000 days. As for the Norman conquest, a time embedded deep in England's history, its year was only 341,000 days before the present. For a million days ago it is necessary to retreat to 737 BC. On the other hand if millions are set against millions the number shrinks decisively. If a supposedly benevolent politician provides, say, £50 million for the gain of British people this means a mere 90 pence for each Briton. Planet Earth lies 93 million miles from the sun. Consequently the radiated light, which hurries along towards us at 186,000 miles a second, still takes 8.3 minutes to make the journey from sun to Earth. As for Concorde, which flies at an impressive and supersonic 1,400 miles an hour, its flight time from Earth to sun would be seven and a half years. Sound travelling that same distance would take twice as long.

Textbooks often state that our sun is an ordinary star, much like the zillions of other stars in the universe. Recent work suggests that our sun is exceptional, being among the most massive 10% of nearby stars. It is also a single star rather than a multiple system, the commoner arrangement. Our personal sun has more heavy elements than others of its age and kind, and much less variation in brightness. Possibly more importantly – for the generation of life on its planet Earth – the sun's orbit is less elliptical than those of other stars. Being more circular this 'prevents it from plunging into the inner galaxy' where life-threatening supernovae are more common, claims Guillermo Gonzales, of Seattle's University of Washington. He believes there are far fewer stars out there which might be suitable as heat and light sources for life to develop on nearby planets.

It should be mentioned that a willingness exists, among some, to believe that voyagers are arriving here from realms far beyond our solar system. These believers may not have sufficiently absorbed the great distances involved, the millions and billions of miles. Astronomers presume that many stars possess suitably attendant planets, as with our local arrangement, and these other worlds may therefore possess life forms, with some of these forms technologically our superiors – by, perhaps, many million years. They will be able to travel speedily, but the distances are still huge. The nearest star to our own is Proxima Centauri which lies four light years away, namely 24 million million miles. Our own galaxy, the Milky Way, contains about 400,000 million stars and is 100,000 light years across, or 2.4 million million million miles. The thought of some alien individuals bothering to travel such vast journeys to reach us, and then perhaps burn a little grass, streak intriguingly across the sky, appear briefly in their saucer-shaped containers, or flatten some English corn before departing, is difficult to contemplate for most, but not for all, of us.

As for the existence of other planets, suitable for some form of alien life, they have been presumed. If the chances of a star possessing a solar system such as ours, with one-ninth of its circling planets suitable for life, are assessed even at a million to one against there are all those billions of stars out there, even within our single galaxy, to make habitable planets a virtual certainty. Evidence for such planets, as against guesswork and probability, arises from the fact that various stars wobble in cyclical fashion. It is therefore supposed – by most astronomers, if not all – that invisible planets are causing these perturbations. About one planet a month is currently being 'discovered', unseen for its darkness but presumed for its effect upon a star. There may be other reasons for the wobbling but these, whatever they are, would be yet more remarkable than the postulation of a planet. As one astronomer put it: 'If something quacks like a duck, and walks like a duck, it probably is a duck. These suns wobble as if a planet is nearby; so there probably is.'

The first extrasolar planet was detected – from the wobbling – in 1995, and 28 more were found before the end of the century. Then, in December 1999, it was announced from the University of St Andrews, Scotland, that light had been observed from a planet in another solar system. This was a blue-green glint discovered by using spectral analysis with a telescope in the Canary Islands to separate a planet's lesser light from the tremendous glare of its parent star. The star in question, Tau Boötis in the constellation Boötes, lies 51 light years from Earth, or 300 million million miles. The planet, thought to be 3.9 times more massive than Jupiter, orbits its sun every 3.3 days, is 20 times closer to its parent body than Earth is to its sun, and is therefore one of the hottest planets so far known or presumed. Its sighting means that a 'revolution in planetary science' is being predicted, said an enthusiast for this discovery.

One emphatic wonder nearer to home, bearing in mind the remoteness between our life-giving sun and its dependent planet Earth, is that we receive so much heat and light from our single star. Earth is but a speck from this sun's viewpoint, much like a football seen from two miles. The sun radiates heat in every direction and we experience a mere 0.00000000046 of its energy output. Putting that figure another way round it means that we receive 460 units of energy for every trillion created and radiated by the sun. The other planets in our solar system receive different quantities, with Mercury – nearer the sun but smaller than Earth – obtaining a little less than us. As for distant Pluto, even smaller than Mercury and very much further from the sun, it is blessed with about one 40,000th of Earth's supply.

Not all of the energy which arrives at the upper levels of our atmosphere reaches the Earth's surface. Some is absorbed by those upper layers and some is reflected back to space. The calculation has been made that, if Earth was not a sphere but a flat disk facing the sun, it would receive about 1,145 watts of energy for each square yard of its surface. As it is spherical, with very little of its surface actually facing the sun directly, and with half of its surface always facing in the other direction (during its hours of darkness), the average quantity of energy received by each square yard falls to 286 watts. The atmosphere radiates about 6% of its received energy back to space, and Earth's surface reflects a further 10%. Therefore the received quantity is 84% of 286, namely 240 watts per square yard. By way of comparison a colour television set consumes and therefore radiates about 240 watts of power. So, of course, do four 60-watt light bulbs. Those individuals who often keep a young animal warm under a single light bulb – such as vets, zoo keepers and concerned owners of ailing pets – are probably providing heat roughly equivalent to the sun's warmth at Earth's surface on a bright and sunny day.

As Earth receives this quantity of heat, and as its temperature stays reasonably constant, it follows that 240 watts per square yard must also be lost. Warm surfaces radiate more heat than cool surfaces, and it has been assessed (by the Stefan-Boltzmann constant, for those who comprehend it) that the Earth's surface temperature should be minus six degrees Centigrade if it were truly balancing the amount of heat received. As it is much warmer, being about 15 degrees C., something else must be at work to explain this divergence. That something else merits an entire chapter to itself, being the effect of certain gases in Earth's atmosphere. The word greenhouse has achieved much fame, owing to the fact that human activity may be promoting this effect, but see Chapter 13 for many pages devoted to this single topic.

Back now to the huge quantities of energy which the sun produces. In doing so it converts four million tons of its mass per second while transforming hydrogen into helium, but the resulting radiation is not constant. Our star oscillates, changing its diameter every 2 hours 40 minutes, and it also produces sudden bursts of extra activity. Great bubbles of gas abruptly appear, perhaps 20,000 miles across, before shooting into space. One recent explosion was likened to the force of 100 million hydrogen bombs. Sun spots – darker and cooler areas, by about 1,000 degrees, hence the relative blackness – are frequently visible. (More complicatedly, and according to recent definitions, they are dark either because of 'the partial suppression by the sunspot magnetic fields of convective energy transport from the underlying layers' or by 'the removal of energy from the sunspot by enhanced hydromagnetic wave radiation'.) Such spots tend to be either most or least numerous in 11-year cycles, and the sun itself has an 11-year cycle, the quantity of its emitted energy varying by 0.1%.

As something else for us to worry about, far more powerful than flares reaching out for 20,000 miles, there are solar explosions known as super-flares. These have been studied on other sun-like stars in our galaxy, and are formidable. If one of these outbursts were to happen on our own sun the energies involved would be about 10,000 times as powerful as the large and occasional flares which Earth experiences, such as the one in 1989 which famously knocked out a power grid in Quebec. (It was not a good time to lose electricity, and for the furnaces of six million Canadians to be switched off. The outside temperature on that chilly night of 13 March was minus 15 degrees C.) The super form of flares make their suns dramatically brighter, sometimes for hours, sometimes for days. If one were to shoot from Earth's sun, according to an astronomer from Yale, life on Earth would survive, but 'half the ozone layer would immediately be destroyed' and 'all our satellite fleet would probably be lost'. So why has this not happened? Either our star is extremely stable or, less comfortingly, it happens to be stable right now, but may not always be so.

Sunlight is electromagnetic energy and comes to us by electromagnetic waves. Within the spectrum of sunlight are:

Ultraviolet radiation (UV), which we humans cannot see;

Visible light, which we can see; and

Infrared radiation, which we also cannot see, and which is our main
source of heat.

Ultraviolet radiation, however invisible, is certainly of concern, with this
anxiety likely to increase. Our efficient eyes make it easy for us to believe
that we see all light, but we cannot detect the form of short-wave radiation
lying between the 'visible' spectrum and X-rays. The fact that certain
other creatures can see this different band of light was beautifully
illustrated in 1879 by Sir John Lubbock when he was working in the
basement of London's Royal Institution. Knowing that ants rescue their
pupae from unwelcome light he scattered many of these developing
offspring over a floor bathed in light from a quartz prism. At the red end
of this spectrum all the pupae were left untouched, but at the violet end,
and particularly beyond that violet end, the inert pupae were promptly
rescued and placed in darkness. The insects therefore proved they did not
see redness as a form of light but, contrary to human vision, could see the
ultraviolet.

The fact of the sun's surface being so hot means that 9% of solar
radiation lies in the ultraviolet range, as against 45% in the visible
spectrum. As everyone knows who has ever put a poker in the fire the kind
of light emitted by a hot source is dependent upon its temperature. Pokers
can be red hot, or white-hot if warm enough, or even dark when giving out
no visible light but still far too hot to touch. Earth, whose average surface
temperature is very much less, gives out low intensity radiation mainly in
the infrared range at quite the spectrum's opposite end.

In modest doses the sun's emission of UV is beneficial, being important
for the synthesis of vitamin D in humans. (This vitamin, like other
vitamins, cannot ordinarily be created within the human body; but, unlike
the others, can be made in the presence of sunlight.) Excessive UV dosage
can lead, as the World Health Organisation phrased it, to 'various skin
cancers, accelerated skin ageing, cataract and other eye problems', as well
as 'adversely affecting people's ability to resist infectious diseases'. UNEP,
the United Nations Environment Programme, has estimated that more
than two million non-melanoma skin cancers and 200,000 malignant
melanomas occur globally each year. WHO produces a Global Solar UV
Index, ranging from Low (1–2), Moderate (3–4), High (5–6), Very High
(7–8), and Extreme (greater than 9).

Near the equator at midday in summertime the values can range up to

20. Thinning of the ozone layer in the stratosphere, a shrinkage discovered in the mid-1980s, has caused concern mainly because this form of oxygen protects the Earth's surface, and Earth's inhabitants, from excessive UV radiation. UNEP has stated that, in the event of a 10% decrease in stratospheric ozone, there will be an additional 300,000 non-melanoma and 4,500 melanoma skin cancers worldwide. Humans without much skin pigmentation are more at risk from both forms of cancer. Increasing attention will certainly be paid to the UV index so long as Earth's protective ozone continues to be depleted. The global loss of ozone has been estimated at about 4% a decade throughout the past 20 years, and no one expects this lessening to be halted for many more decades. WHO asserts that the UV index should form part of all weather forecasting.

It is a wonderful fluke that we have eclipses, with the sun's diameter being 400 times greater than the moon's, this fact so happily coincident with the sun being 400 times further away than is the moon. It used to be terrifying for our forefathers, when the precious and much revered sun was obliterated for no known reason, save for the presumed whims of gods, but the event of an eclipse is still extraordinary, even in a world often blasé about natural phenomena.

The Chinese are believed to have been the first, as with so much, to record one of these events when 'the day dawned twice at Zheng'. This inscription on bamboo is thought to have been written about three millennia ago, and the only possible eclipse occasion at that time was 21 April 899 BC. There is dispute about this inscription's authenticity, partly because the concept of a double dawn is written nowhere else. At least, it was not written anywhere else during those early years. However there is a phenomenon known as a false dawn, a brightening of the sky a couple of hours before the true dawn. There are several explanations for this odd, and rare, occurrence, but not one of them suggests it is connected with eclipses. By 708 BC the Chinese astronomers knew exactly what was happening, and that the moon had positioned itself between Earth and its sun. Perhaps they had divined this explanation in the interim, between 899 and 708, or unwritten legend had come up with the poetic description of a day beginning twice?

The solar eclipse of 11 August 1999, already mentioned (in Chapter 1) and which raced its totality across the world from dawn in the West Indies to dusk in the Bay of Bengal, was allegedly observed by two billion people, much to their amusement, excitement, astonishment, enchantment, bewilderment, or all five more or less at once. *The Times*, of London, described this mega-human reaction as a giant Mexican wave travelling across the planet at more or less the speed of sound, with those two billion people standing up, leaving their daily tasks, pausing awhile before the heavenly spectacle, and all being forcibly reminded that their actual home is no less lonely and spherical than those two nearest neighbours of a sun and a moon, these three objects temporarily in line to provide a stunning sight.

'Perfect' eclipses, as seen from Earth, and akin to 11 August 1999, will only be possible for the next 150 million years. Similarly, despite Earth's extreme antiquity, they have only been possible for the past 150 million years. Tidal effects are causing the moon to recede, and we now happen to be living during a band of 300 million years when the moon's distance from Earth permits eclipse perfection. We on our planet are also singularly fortunate in that a perfect eclipse is not possible with the 64 other moons circling their attendant planets in the solar system, almost all of these natural satellites orbiting the giant outer planets of Jupiter, Saturn, Uranus and Neptune.

Even our existence here on Earth, and of life in general, is believed by some to have been dependent on the nearby presence of an unusually large moon. This impressive neighbour, with a diameter one quarter of Earth's, prevents our planet from extreme wobbling on its axis which would lead to massive swings of climate, such as those which exist on Mars. It also causes tides which may have assisted in the evolutionary process, bathing living forms rhythmically first in water and then in air.

As for the distant future, long after the time when eclipses are no longer possible, changes to the sun will become distinctly more important. Because our sun is getting warmer by about 1% every 100 million years there will be more water evaporated from our planet's oceans. This increase in water vapour will serve as a greenhouse gas, thus causing Earth to become hotter still and therefore yet more water to be evaporated. About a billion years from now the terrestrial temperature will be 60 degrees, and 10% of the atmosphere will be water. This baking situation will continue for some 3–5 billion years until the sun exhausts its hydrogen supply and becomes a red giant – as seen and defined by astronomers for objects all over the heavens. By that time our star will have expanded prodigiously, perhaps to 1,500 times its present size, and will have engulfed Mercury and Venus. From Earth's viewpoint the sun will fill most of the sky. Although our planet will then be severely cooked, to say the least, the outer planets of our solar system will have been warmed to more reasonable levels than they currently experience. Life may even begin out there, serving as replacement for its extinguishment on Earth.

As for Earth's inhabitants, assuming they have not earlier destroyed themselves in some fearful conflict or mismanagement, they will have had plenty of time to see disaster looming. Perhaps some of them will be able to leap-frog to the outer planets, as and when these become suitable for life. Or perhaps they will even travel to some other solar system which possesses a younger and therefore more acceptable source of light and heat.

If Earth were a two-dimensional circle facing the sun its climate would be reasonably uniform; but the fact of being spherical means more heat hitting the equator than the poles. The further fact that Earth's axis is canted at an angle to the sun means that everywhere receives varying levels of radiation during Earth's annual cycle around the sun. As third crucial fact the Earth's orbit is elliptical, causing Earth's total of received radiation to alter as the year progresses. A final astronomical point, crucial to climate, is that Earth itself is spinning. Each point on its surface is either receiving sunshine or it is not, a daily (or annual – at the poles) rhythm which has existed since the beginning. This spinning is slowing down, with more days per year when reptiles were pre-eminent and even more in all the earlier ages.

Hence, in brief, our weather. More heat hits Earth's central belt than its poles. This belt of greatest heat moves north and south as the Earth's tilt favours one or other hemisphere. Earth's elliptical orbit causes the planet as a whole to receive more heat for half of each year than for the other half. No wonder, therefore, that there is both climatic confusion from the conflicting conditions and also an assortment of generalities.

First, therefore, to a few of those generalities. Heated air rising in the hotter regions, the portions of Earth most subject to solar radiation, causes cooler air to travel from nearer the poles as replacement. The rotation of the Earth alters these air movements, deflecting them to travel either east or west, and causing other winds to take their place. Globally there are easterlies – the trade winds – blowing between 30 degrees North and 30 degrees South. On either side of this belt, and between 30 N. and 60 N. in the northern hemisphere (or between 30 S. and 60 S. in the south), there are westerlies – which is why most of British weather comes from the west. At the boundaries between these sets of winds blowing from east and west, and also at the equator, there are often windless conditions, the doldrums. Such areas, notably in the north, were sometimes called the horse latitudes, when slow sailing caused thirsty horses to be tossed overboard to economise on water. North of 60 N. and south of 60 S. the westerlies are replaced by easterlies, with the poles themselves usually regions of high pressure, and therefore calm.

So much for the broad picture but the encircling atmosphere, becoming less and less dense with altitude, is itself influential. Aircraft flying at 30,000 feet, the sort of height exploited by the long-haul airlines, have over three-quarters of Earth's atmosphere lower than their cruising altitude. This is as difficult for most passengers, intellectually burdened by tea-trays and in-flight movies, to comprehend as either the plane's air speed being a reported 600 miles an hour or the outside temperature beyond those plastic windows being a chilly minus 40. The halfway atmospheric point, with half of its air above and half below, occurs at about 17,000 feet, this height only slightly greater than Europe's highest point. With increase in altitude there is a tendency for winds to become more westerly, and aircraft flights from west to east are generally speedier than flights the other way. By 40,000 feet (7.5 miles) wind direction is almost always from the west, save – on occasion – near the equator. Very much higher, above 20 miles or so, there is a reversal, with winds usually flowing from east to west.

Jet streams were first detected and named by American pilots, although they had been suspected earlier by meteorologists intrigued by fast-moving upper clouds. The airmen did not understand why their estimated ground speeds and times of arrival were proving to be so faulty, with planes reaching their destinations either long before or after expectation. Jet streams are most impressive during each hemisphere's winter season, when very cold air sweeping from the poles forces its way southwards in the northern hemisphere, and vice versa in the south. They behave so vigorously that winds of 200 miles an hour can result, these travelling from west to east. Like orange pips squeezed between thumbs and fingers the cold air masses meet warmer air, and something has to give. Balloonists attempting to circumnavigate the Earth, and even succeeding on one occasion, or merely to travel between continents, have exploited these speedy air flows, it being more profitable to drift at 200 miles an hour than at one tenth of such a speed at lower altitudes.

With increasing distance from the equator the sun's warmth has to travel at an increasing angle through more atmosphere before encountering the ground, with the poles receiving only 18% of what they might expect if there were no atmosphere. Warmth is dissipated by having to travel a greater distance through the air.

Such generalities are possible for some fundamental reasons: Earth is a planet; it has an atmosphere; it is revolving; and it is orbiting the sun in a slight ellipse. The general points arising from these basic reasons remain broadly true despite the extra fact that Earth's surface is not uniform. Its atmosphere is fairly consistent, with ground level pressure only varying normally between 950 and 1,050 millibars, and with its contained water vapour gradually decreasing with altitude, but everything at the ground on which that atmosphere rests is extremely varied.

Some of it is mountainous, reaching, at its maximum, 5.5 miles above sea level. Mountains cover 9% of Earth's surface, with the world's greatest range, containing all peaks above 8,000 metres/26,246 feet, more than 3,000 miles long. There are also deserts, savagely hot by day and frequently freezing at night. They cover seven million square miles, 12% of the land surface, and exist mainly in the northern hemisphere. Conversely there is ice, with 90% of it resting on the 5.5 million square miles of Antarctica where average ice thickness is 7,500 feet. That continent doubles its colossal area every winter as the nearby oceans freeze.

Then there is forest, entirely different to ice, to desert or to mountain range, and there is tundra, different yet again, being without a tree, extremely flat, pocked with lakes, and covering five million square miles, or 9% of all the land. This particular feature is therefore bigger than Europe or the United States, and a thousand times the size of England. The so-named semi-desert, less arid than true desert but still wholly unfit for agriculture, covers a further 4% of Earth's land surface.

The greatest surface difference of all, greater than the several varieties which create each continent, occurs when land becomes water. Earth's oceans cover 140 million square miles, or 70.8% of the planet's surface area, with actual land, including lakes, rivers and other forms of wetness, less than a third. (As Arthur C. Clarke so aptly wrote: 'How inappropriate to call this planet Earth when it is quite clearly Ocean'.) Sea temperature only varies between minus 2 degrees C. and plus 29 degrees. Its warmth is far more consistent than land where temperatures can fall to minus 88 degrees, as in Antarctica, or rise to plus 58 degrees as in Mexico, this range of 146 degrees being 46 Centigrade degrees greater than the temperature difference between ice and boiling water. Even single areas, as in eastern Siberia, can range in one year between minus 60 degrees and plus 45, also greater than between water's melting and boiling points. Britain is thrown into confusion should the temperature drop to minus 5 or rise to 27 – 'Icy horror grips East Anglia' or 'Phew, what a scorcher!' Browning, Montana, once recorded a shift of 37.8 degrees in a single day, this being greater than the British astonishments in a year.

There is relatively more water in the southern hemisphere, a further fact to disturb any climatic balance between the two hemispheres which might otherwise exist. As all the major surfaces, whether ice, lake, swamp, desert, semi-desert, forest, savannah, tundra, mountain or ocean, react differently to the solar radiation which comes their way, it is plainly of significance to climatic variation how much of each type exists around the world. With water taking longer than land to cool, and then longer to heat, each northern summertime does not cause northern oceans to be warmest until August or so. Deserts absorb heat quickly but speedily lose it at night. Forests, swamp and so forth are intermediate between the two extremes of ocean and of desert, one being so slow in its response to heat and the other so swift.

Human beings have also caused different land surfaces of their own, such as their unnatural fields or even more unnatural buildings; but people, for all their numbers and all their influence, only live and grow crops on some 12% of Earth's land surface, or 4% of the planet's surface as a whole. All the various crops absorb or dispel solar heat differently, and these differences alter as they grow. Towns and villages are different yet again. A house maintains solar radiation best via its walls and much less with its roof, the vertical surfaces bouncing heat from wall to wall, and tending to retain it rather than dissipate it elsewhere.

Coupled with the major surface zones, each occupying several percentage points of Earth's total area, are various lesser regions. These too can be formidable in their effects upon climate, upon wind and rain and warmth. A single peak like Tanzania's Kilimanjaro is a pimple on the map of Africa, but its circular base is 45 miles in diameter. Therefore the area consumed by this one volcano is 1,600 square miles (bigger than England's Kent and much bigger than the state of Rhode Island). Ten per cent of Spain is bare rock, very different from soil as an absorber and radiator of heat. There are lakes and there are cliffs, there are rivers and roads, swamps and hills, dry bits, wet bits, and an infinite variety of surface coverings, all of which have different local effects which then affect the bigger picture.

Small wonder, therefore, that wind direction and strength, precipitation or its lack, and low pressure regions or high pressure zones are all mightily influenced by our planet's furnishings, whether these are steep, high, low-lying, barren, lush, bushy, dusty, rocky, or part of the great amalgam which is man-made. It is almost more surprising that any rules apply, such as winds travelling anticlockwise in the northern hemisphere around centres of lower pressure, and also anticlockwise in the south but around the highs. Rules exist because the planet is so huge in relation to alterations on its surface.

Everest is big but its 29,000 ft height is dwarfed by Earth's 8,000 mile diameter, being equivalent to a wrinkle one yard high in a distance of one mile. The whole Himalayan range from Afghanistan to China is extensive, but that 3,000-mile length is only one-eighth of Earth's circumference. As for the oceans, although irrelevant to a section on atmosphere, their inverse wrinkle is also modest relative to Earth itself. They have been likened to puddles, with the deepest trenches of the deepest oceans only slightly further from sea level than Everest is higher. Average ocean depth is 2.3 miles. This is more or less equivalent to the Tibetan plateau, if depth versus height is compared.

These surface creases and corrugations, the unevennesses of high bits and low bits, are of critical importance to weather because the atmosphere itself is also of modest size in relation to the Earth which it surrounds. The so-called microclimates can reinforce this point. From an insect's viewpoint it is important to be deep within grass or on top of it, beneath a leaf or above it, and either in or near a puddle. There are only two insect species, for example, living permanently on the Antarctic mainland. These both live within such slender vegetation that exists, and neither possesses wings. In that windy, freezing world it would be fatal for them to fly – anywhere – from some relatively cosy microhabitat. On a bigger scale every farmer knows that one field is best for crops, perhaps near that belt of trees, by the stone wall, furthest from the stream, and where sunshine is greatest with frost least likely. Local conditions on any scale can make or mar success.

House builders in earlier days, knowing of flood levels in wet years, of frost traps in cold ones, and of wind-chill every year, did not build dwellings in unsatisfactory spots. Today's builders, knowing that floods can be diverted, that central heating is effective, and windows keep out wind, can be over-casual, or just greedy or ignorant, in siting homes. The purchasers are then amazed by high fuel bills, by tremendous draughts, and on occasion by water entering the door. Animals can be far more adept in selecting appropriate spots to live and rear their young.

Light at the bottom of a forest may only be one per cent of the quantity at its canopy. Ordinary grass is similar, with conditions varying according to height above the ground. Wind speed in long grass, assuming a soft breeze, is markedly reduced with decreasing altitude. A breeze of 44 inches per second/2.5 miles an hour at 16 inches above the ground becomes 24 ins/sec at 12 inches, 10 ins/sec at 8 inches, and virtually nil at 4 inches or lower. Similarly, and on such a day, if the temperature is 20 degrees C. at 2 feet, it is 22 degrees at 12 inches, 21 degrees at 8 inches, 19 degrees at 4 inches, and 17 degrees at ground level. Water vapour also varies with height being, on the same kind of day, perhaps 18 millibars at ground level and rising to 15 millibars at 16 inches and above. With wind speed either still or 2.5 miles an hour, with warmth shifting from 17 to 22 degrees, and water vapour changing from 18 to 15 millibars, the alterations are substantial for delicate life forms whose entire world is provided by that grass. In one of those environments they will thrive; in the others they may die. Such creatures can choose between life or death by climbing a blade of grass.

Similarly there is a tree line on our planet. South of it (in the northern hemisphere) trees can exist and grow; above it they cannot. This line is almost identical to the ten degree C. isotherm, which joins all points whose warmest month in the year averages ten degrees. The two lines do not match precisely, it being more critical for trees that three months have temperatures averaging six degrees, or higher, than one month averaging ten degrees.

Major events can also be hugely influential. A single volcanic eruption may eject billions of tons of dust into the atmosphere, altering the climate worldwide. A meteoric impact can be yet more devastating. It is generally accepted that the near-total demise of reptilian and much other life 65 million years ago was caused by a massive arrival from space which landed, it is thought, by southern Mexico. Even the gentle moon, so benign in its appearance, has power to alter terrestrial events. The tides are well known, twice daily alterations of sea level which are at their most severe when sun and moon are in alliance, but the atmosphere is also influenced by the moon's positioning.

To sum up this introduction to terrestrial weather – Earth's principal heat and light source itself varies in its radiations. Earth varies in its distance from this source, and spins with its axis at 66.5 degrees, thus changing each location's ration of radiation according to the time of year. Each of Earth's revolutions lasts for 24 hours, with slightly more than 365 of them in every peregrination around the sun. Earth is spherical, if not quite perfectly, and its atmosphere and shape both affect the quantities of light and heat arriving on its surface. The ground level atmosphere varies in its density, ranging from extreme lows of 880 or so to extreme highs of 1,080, a difference of 20%.

All of this affects a surface which is also extremely varied – in texture, in retention of temperature, and in being liquid or solid. Loss of heat every night varies considerably, from swift as with deserts to slow as with oceans.

Air movement is no less variable, in part dictated by Earth's rotation, in part by pressure differentials, and in further part by all the features of the land, such as mountain ranges big or small, such as forests, such as everything which exists. Winds alter with altitude, with temperature, and with the seasons. These may or may not carry much water vapour, and therefore may or may not form clouds, either near Earth's surface or far from it. The clouds themselves, whether actively producing rain or not, undoubtedly adjust heat's arrival from the sun, either blocking it or permitting varying degrees of passage. Volcanoes happen. Hurricanes happen, either catastrophically or more benignly. There are dry years, wet years, torrential years and drought years, which themselves cause further events. Everything impinges upon everything else, continually, remorselessly, confusingly. On no two days, it can safely be assumed, have all the climatic conditions on Earth ever been the same.

And yet we humans, although knowing of the numerous possible perturbations, frequently expect a sameness to prevail. We have our man-made calendar, itself regular and almost identical from year to year. We therefore tend to believe, or wish, that the climate, in every sense, should follow suit. 'I've never known a January like this, not here, not in this village, not since I've lived here.' 'The swallows are a week late, so what's up?' 'Leeds has never known such winds, not in March.' 'Why can't it be like last year with its terrific August?' 'We've had a month's supply of rain in the first week.' 'Already we've had twice as many hurricanes/tornadoes/wildfires/avalanches/ landslides as usual.' 'The reservoirs are half empty which they shouldn't be, at least not in June.' 'It isn't natural.'

Yes, it is. Variation is entirely natural, but the big question today is whether, for once, we may be right in claiming a lack of naturalness. Are we, in part, responsible for any climatic changes extra to all the changes occurring naturally? Just as no two days on our planet have ever been identical so has the current situation never had its precise equal in the past. Global warming, glacial retreat and altering ocean levels have often occurred in different forms and in former times. These changes are still happening. The query, the biggest and most gigantic climatic query of our age, is whether and to what degree the current crop of alterations is being caused by us.

Chapter 4

CLIMATIC ZONES

Generalities – global averages – seasons – warmth for growth – atmosphere – air pressure – troposphere – stratosphere – latitudes 43.5 N. and 62 N. – Stevenson screen

Earth's axis is constant. Earth's orbit is constant, and the sun's rays steadfastly approach each part of Earth with similar intensity. It might therefore be expected that some form of climatic constancy would prevail on Earth, according to the time of day, the time of night, the season of the year, and either the all-important distance of each place from the equator or its nearness to a pole. Instead there is massive variety, as the previous chapter indicated, with the actual latitude seemingly irrelevant at times. One place can be consistently colder than somewhere well to the north of it, or warmer than somewhere far further south.

Verkhoyansk, for example, lying within the Arctic circle and therefore north of northern Iceland, can experience the heat of 38 degrees C. in summertime. Contrarily, and in winter, its thermometers can drop to minus 71 degrees, far colder than is ever reached either on Iceland or at the North Pole. New York City, with cold winters and hot summers, is 11 latitude degrees nearer the equator than London, this British capital tending to have relatively balmy winters and unconvincing summers. Edinburgh is on the same latitude as Moscow, Los Angeles is south of Algiers, and Volgograd (where both armies knew such terrible winters when it was Stalingrad) is on the same parallel as Paris. So are Buenos Aires, Oklahoma, Wilmington, Adelaide and Canberra, all similarly far from the equator and all with great dissimilarities in their climate. None of Africa – not even its temperate tip near Cape Town, cool in winter, warm in summer – is as far from the Earth's central belt as all those cities. Much of the United States, and not just Alaska, has extreme winters often with massive snowfalls, these chilly areas existing well to the south of Britain and also south of France's warmer Brittany. Look at any map showing lines of equal temperature, or equal pressure, or equal cloudiness, or equal sunshine or equal anything, and neither the lines of latitude nor the landmass shapes show much relationship with any of them.

The British anomaly, as well as that of the fringe of western Europe, exists because the famous Gulf Stream brings warm water swirling north-eastwards from the south-western Atlantic. This beneficent current is the reason why, for example, an extraordinary garden of exotic vegetation can thrive at Inverewe, only 65 straight miles south of Scotland's north-western corner at Cape Wrath. That garden's latitude is 58 degrees N., almost that of St Petersburg, but the place never receives more than 2–3 degrees of frost – hence its giant forget-me-nots from the South Pacific, its bamboo from Asia, and its tremendous eucalyptus from Australia.

The Gulf Stream's warmth does not penetrate far into Europe, with the Baltic quite a different place to the waters between Britain and Norway, but the fact of a landmass obstructing ocean currents, of diverting and of blocking them, is not unique to northern Europe. Neither is that of a continent's interior possessing quite a different climate, well away from the benign influence of warm oceans. The distinct properties of land and water combine to play havoc with the climates which might reasonably be expected if distance from equator or pole formed the sole criterion. As for the North Pole it may look a forbidding place, and is hardly welcoming, but its ice sits upon unfrozen water. Ninety degrees North is therefore warmer than if it sat on land. Antarctica is solid land below its icy covering, and Antarctic temperatures are massively colder than those of its Arctic counterpart.

It is difficult making useful generalisations about climatic conditions here on Earth. Winds can be non-existent even for several consecutive days in certain regions, a fact difficult to understand when the sun's warmth and powers of convection vanish so absolutely every night. Winds can also gust at 200 miles an hour when hurricanes or typhoons are in full swing. Airborne moisture can be zero, notably over deserts, or 100% elsewhere. Rainfall can be nothing whatsoever, year after year, notably in northern Chile, or 25 yards in a single year, as in parts of northern India, this amazing figure perhaps more comprehensible than 905 inches or, which meteorologists prefer, 22,990 millimetres. Manchester, famed for wetness, receives 30 inches annually, less than a single yard and a mere 762 millimetres.

That city, strangely for its reputation, receives about the global average for rainfall on land. The oceans receive rather more, about 50 inches a year. Most of the evaporation leading to this rain occurs from the oceans, a formidable 123,000 cubic miles of water per year. A further 17,500 cubic miles arise from the land, and 11,000 cubic miles of land's rainfall departs as run-off via the world's rivers. The land's rainfall, the steadfast blessing on which we all rely, is therefore 28,500 cubic miles. It is easy to wonder, with water the most basic resource of all and increasingly utilised by an increasing population, whether 38% of this libation will forever be allowed to flow into the sea. Virtually none of the river Nile now reaches the Mediterranean, and such denial may become the pattern. The Amazon's flow, pouring into the Atlantic, is some 11 million cubic feet a *second*.

Once evaporation has occurred, and water has returned to the atmosphere, it only stays there on average for 8.2 days. This brevity is one reason why weather forecasting, coupled with the predominant human wish to know whether rain will fall or not, is so tricky, save for the immediate future. At any one time about 0.001% of the world's water is in the atmosphere, as against 96.5% in the oceans and 1.7% as polar ice. Rivers contain even less than the air by possessing only 0.0002%, although this quantity is 0.006% of the world's fresh water. Polar ice represents 68.6% of the fresh water total, and a further 30.1% lies far below our feet, even though we are more conscious of, and certainly reliant upon for agriculture, the 0.5% which is present in the soil.

As for clouds, which may or may not lead to rain, this covering can be unknown in various arid areas or perpetual in others. The phenomenon of evaporation, leading to more moisture in the air, may be swift, when the air is both hot and dry, or extremely laggardly when the nearby air is either cold or already damp. But, somehow or other, and somewhere or other although mainly over the oceans, enough moisture is gathered to produce the annual global rainfall of 140,000 cubic miles of water, of which about one quarter arrives so crucially on land.

The presence of snow or ice at spots around our globe may be quite unknown or occasional, an annual happening or quietly permanent. Ground surface temperature may be either far above freezing or never achieving melting point. As for the critical and life-providing soil, that covering may be dusty, crumbly, firm or rock-hard as with permafrost.

The several astronomical seasons per year, from equinox to solstice and so on, can either be extreme, and tremendously influential on the climate, or so similar they are scarcely noticeable. Ocean Island in the Pacific, lying near the equator, experiences the same mean monthly temperature of 21 degrees C. throughout each year. Average warmth in England and Wales oscillates between winter and summer by 11 degrees C., shifting from 5 to 16 degrees, with the local inhabitants expressing concern when days are hot in summer or cold in winter. This yearly alteration of average monthly temperature, so different from Ocean Island, is some five times greater in eastern Russia.

Human beings are said to be tropical creatures who create their own climate wherever they go. Most living things cannot be so adaptable, having to experience whatever comes their way. Trees live only on one side of a boundary, north of which in the northern hemisphere they cannot exist for lack of warmth. Most British vegetation cannot grow if the temperature is below 5.5 degrees C. In the tropics, with year-round heat, trees can increase their height at almost visible speeds, the tree-weed known as Cecropia achieving 15 feet or so a year, which is half an inch a day. Warmth is important everywhere, but it is the year-long lack of cold around the equator which leads to its tropical exuberance.

The world's tallest trees do not grow by the equator, as might be expected. The real giants are the Redwoods of California and Oregon, well to the north of it, or the many lofty species of Eucalyptus which flourish far to the equator's south in Australasia. That margin of 5.5 degrees is particularly critical in temperate regions where an early spring or delayed autumn significantly increases the period for growth. In Cumbria, a county embracing a hilly portion of north-western England, the growth span of low-lying and near sea-level land is 37 weeks. That quantity is reduced to 27 weeks a mere 1,500 feet up the slopes. At 1,900 feet, or about as high as any fields have ever been constructed in that area, the span is 23 weeks. Such a small growth season, being less than half the year, is inadequate for most crops. They cannot flourish sufficiently in any single year to be of use.

Vegetative growth still occurs well to the north of Cumbria, but even less vigorously. On the tundra, that bleak and flattish vastness north of the tree line where 60% of the land is water although there is hardly any rain, no plant can grow until mid-July and such belated development halts in September when winter chill arrives once more. So what do plants do in that locality, for there are plenty of them and they can look magnificent? One solution has been a spreading of the annual cycle, this lasting for three years instead of the conventional singleton for normal 'annuals'. Grow one year, produce flowers the next, and then make seeds in the third and final year. With warmth, of a sort, only lasting for seven or eight weeks the compromise makes sense.

Basic rules for winds have been mentioned in Chapter 3, but they are guidelines more than law. A prevailing wind pattern may be from the west, as with Britain and much of western Europe, but this region also

experiences winds from east, north and south. If two-fifths of the winds
blow from one direction, and one-fifth from each of the three other
quarters, the major wind direction is said to be prevalent, even though
60% of the wind is arriving from those other areas.

The continents themselves are big disrupters of any general pattern.
Asia, largest landmass of them all, is particularly contrary with a
wintertime clockwise flow of air over the continent becoming anticlockwise
in summer. A general rule for the northern hemisphere is that
anticlockwise winds are cyclonic – they go that way round a region of low
pressure. Therefore stand with your back to the wind, and the low pressure
centre is on your left. The precise opposite is true for northern anticyclones
and also for lows/depressions in the southern hemisphere.

Cold winds, so runs an English adage, are often 'lazy', going 'through
one's body' rather than around. Of course, whatever they may feel like to
a human, they do nothing of the sort. Neither do they go through hills,
mountains or any obstruction. All land projections have tremendous
influence, causing winds to bypass them and regroup on the other side or
over them, thus – probably – causing them to drop their moisture on the
windward face. On each mountain's other, and leeward, side there may in
consequence be a dry spot, a rain shadow, which is probably warmer as
well as drier. This heat will cause air to rise, and this rising will further
disrupt the air flow, much as the rising height of land has already done.
Nothing, in short, is uninfluential.

If there is water in the region, whether as lakes or seas, there will be
further disruption, with water less violent in its temperature change than
the surrounding land. If there is snow or ice this will reflect more heat than
either land or water. Therefore such bright areas will tend to stay cold.
There are also sea breezes and land breezes, these usually occurring on a
daily basis, with each being caused by the relative warmths of sea and land.
Hot air picks up more water than cold air. More water vapour in the air
makes clouds more likely. Clouds discourage sunshine from reaching the
ground, whether the ground is damp forest, dry desert, warm water or
urban landscape. Denser air may lead to regions of lower pressure, and
lower pressure causes winds to flow from high to low cyclonically – anticlockwise – in the north.

Everything is interdependent. The fact that swallows are a week late, or
that a month's average rainfall arrives in that month's first 14 days, should
therefore seem entirely reasonable. It is more extraordinary, bearing all
the cross-currents, the countless disturbances and the conflicting forces in
mind, that anything 'typical' ever happens, that rain does fall more or less

as normal, that deserts do stay dry, that clouds do form where they usually form, and swallows do arrive every year within a few days of their expected date. When small exceptions occur there is a human tendency to blame some unnatural event for what, almost certainly, is a natural fluctuation. Gunpowder and its explosions were once indicted for climatic discrepancies. Witches were certainly blamed, being scapegoats for almost everything. When steam trains arrived they too were considered pertinent, with their puffs of steam so resembling and therefore 'causing' clouds. Then came nuclear detonations, their power so horrendous. Surely explosive forces measured in megatons of TNT-equivalent must be affecting the world's climate, quite apart from their devastation of the surrounding area?

It is easy, from our standpoint in the 21st century, to mock earlier attributions, partly because they have been replaced by an even bigger concern. Is everything we do, everything that fuels our current livelihoods, to blame for the climatic changes which we, with better equipment than previously available, are now able to detect? And are the man-made effects sufficiently powerful to amend the sum total of events which our planet, via all its influences, inflicts upon itself?

Climate is essentially an atmospheric happening. The sun's rays are at the root, heating land and sea wherever they fall, and it is the consequent disturbance of both the atmosphere and the oceans which moves warmth or coldness from one place to another. The water vapour picked up and carried by that atmosphere then leads to rain or snow. Confusing everything is the effect of Earth's gyration upon its covering of air, that air which permits us to live our lives, and which steadfastly moves from one place to another, creating climates on the way.

Dry air at sea level contains nitrogen, 78.09% by volume, and, to a lesser extent, oxygen, at 20.95%. The remainder largely consists of the inert gas argon, 0.93% and, to a much lesser degree, carbon dioxide, 0.03%, the gas whose rising levels have been causing such disquiet. Very small amounts – less than one part in 10,000 – of hydrogen, neon, helium, krypton, ozone and xenon are also present, the last two in microscopic quantities. None of these gases weighs much, relative to the Earth itself, but they do have weight. Hydrogen and helium, called lifting gases because they buoyantly support weather balloons and all lighter-than-air devices, achieve these roles via their lightness relative to air and not by their absence of weight.

For the planet as a whole the mass of its all-encircling atmosphere is 5,250 million million tons, and that is without the extra of any contained water vapour. Putting this another way round, the weight of the atmosphere in any place is equal to the weight of the top 33 feet of water beneath it because, as scuba divers know, the pressure at that depth equals one atmosphere. The normal method for expressing air density is to say that ground level pressure is 14.7 pounds per square inch, or that the atmosphere at ground level can support a column of water 33.99 feet high, or a 760 millimetre column of mercury. With air so much lighter than everything else with which we come into contact, such as the ground beneath us or our actual bodies, it is easy to believe that air weighs nothing. Strong winds will force the proper truth upon us, particularly if uprooting trees or demolishing our homes.

It is the movement of that tremendous quantity of air, its 5,200 million million tons which, in general, creates our weather. Have, for example, a few hundred million tons just come from the Arctic? And are a few hundred million more being forced to ascend the Himalayas, their rise causing cooling and therefore the dropping of a few hundred thousand tons of water upon those areas down below, which are subjected to a foot or two of rain each day of the monsoon? And might some future weather forecaster state that 750 million tons of cold air at 5 degrees is forcing its way south-eastwards under 1,000 million tons of warmer air at 11 degrees, thereby creating a weather 'front' for us to experience?

A personal experience once helped a group of us, and possibly others afloat that day, to appreciate the tremendous quantities and forces involved when a gentle breeze is flowing across the countryside. We were in a balloon. Our all-up weight of people, basket, envelope and ballast was one ton. Our buoyant gas had therefore displaced slightly more than a ton of air, and yet we were a minuscule part of the sky in which we flew. There were other balloons for us to see, similar specks beneath the clouds, all similar displacements of one ton. The quantity of air in which we were travelling, carried along as part of it, was therefore tremendous in its weight. What we could visualise from our vantage point, but of course not see with air invisible, were millions and millions of tons of atmosphere, every part of it moving at our own leisurely pace of some 15 miles an hour.

It was a sobering comprehension, and quite astonishing. Earth's atmospheric blanket is not some kind of gossamer, insubstantial, vacuous and immaterial. Our most excellent canopy, the air, is a gaseous covering in excess of five thousand million million tons. Even if the total quantity is reduced, for greater understanding, to the amount above one square surface mile, the sort of area we can better understand, its weight of atmosphere is 25 million tons. At sea level that tonnage is therefore bearing down upon each square mile of our Earth. Even one human being, when standing and occupying about one square foot of surface area, has nine-tenths of a ton of atmosphere above either him or her. Such aerial weights are impressive but can be humbled by comparison with all the solid Earth beneath our feet which, in total, is about a million times heavier than its encirclement of air.

If air was of uniform density, and if its sea level mass remained constant with altitude, it would form a layer five miles thick. Instead its density decreases with height, and does so unevenly with increasing altitude. Even at 200 miles over the Earth, at a sputnik's height above the ground, there is still sufficient gas to act as drag on any orbiting space vehicle which, eventually, will cause it to come hurtling back to Earth. As about half the atmosphere is below 17,000 feet, the pressure at that altitude has fallen from 1,000 millibars, the theoretical sea level pressure, to 500 millibars. At 36,000 feet, nearly 7 miles, the pressure is 250 millibars, and it continues to fall with increasing height. At 12 miles it is about 40 mbs, at 30 miles about 1.5 mbs, and at 60 miles only 0.003 mbs. As water vapour is highly relevant to weather, and as air above 36,000 feet contains very little, events above the altitude where the pressure is lower than 250 millibars are therefore less significant.

Temperature also drops off with height. Early mountaineers noted a three-degree Fahrenheit drop for every thousand feet, a fact confirmed and extended by scientific balloonists in the 1860s who were able to rise even higher than the mountaineers. Cooling with height is only partly explained by the sun's warmth passing through the atmosphere and heating up the ground, with distance from that warm ground being highly relevant. It is also explained by air expanding as it rises, having been warmed by the ground. The resultant cooling is equivalent to the gain in heat when air is compressed, as every child first learns when pumping up cycle tyres. If the air carries water vapour, as it is almost bound to do, it will also lose heat when that vapour condenses to form clouds. Saturated air will therefore cool at a slower rate than dry air, but will still cool as it rises. The net result is generally, as those mountaineers already knew, a temperature reduction of 1.6 degrees Centigrade for every 1,000 feet, or about 1 degree C. for each 200-metre rise.

Cooling does not occur when an inversion is operating, when upper air is warmer than lower air. This contradictory state can exist when, notably on a clear night, the ground has cooled very speedily and has also cooled the nearby air. Or when cold air from higher up is descending rapidly, and punches through the warmer air beneath it. Or when a mass of warmer air is forced over a mass of cooler air. Once again, as with the myriad influences which alter wind strengths and directions horizontally, there are conflicting influences which are acting vertically. There are still general rules, such as air cooling with height, and these do operate – save on those occasions when they do not.

The troposphere is defined as that portion of the atmosphere where air normally cools with height, and this region ends with the tropopause about seven miles above sea level. Higher than the tropopause, and continuing upwards, there is the stratosphere. This is concluded at a height of some 30 miles by the stratopause and, higher still, lies the mesosphere which itself is surmounted by the ionosphere.

These various layers might have been better named, their titles not being instantly comprehensible. Tropo is from the Greek for a turning. Strato is not from the Greek for army, such as stratocracy, but from the Latin for a covering. Pause has nothing to do with hesitation, being from the Greek for stop (as with menopause). Meso means middle, and iono is from ion, the term for an electrically charged particle. Earth's ionosphere does indeed contain electrical layers, such as the Appleton and Heaviside, but that second title has nothing to do with heaviness, as might be construed. It was chosen to honour a gentleman who had that name.

Earth's tropo*sphere*, in which we all reside, is indeed spherical in that it surrounds the planet, but it is not of an even thickness. Near the poles, where the sun's heat and therefore convection are at their weakest, it ends at about 26,000 feet/5 miles. Near the equator, where incoming heat is greatest and thunder-clouds can power up to 9 miles above the land, the troposphere extends above 50,000 feet. Therefore, as heat does diminish with altitude throughout the troposphere, its coolest temperatures are above the equator rather than the poles, these being about minus 80 degrees as against minus 40 in the polar areas.

All of Earth's weather, its highs and lows, its rainfall, its fog, its cloudiness and every other variant, occurs within its troposphere. The pressure drop between tropopause and stratopause changes from about 200 millibars to about one. There is hardly any moisture in that huge region, and therefore hardly any cloud. This stratospheric zone, so much bigger than the troposphere, exists from about 8 to about 30 miles above the Earth, and is therefore higher than any of us can reach, save for Concorde fliers when they are cruising and for astronauts when going up or coming down. Within the stratosphere there is a gradual warming rather than a continuation of the cooling within the troposphere. At the stratopause, upper limit of the stratosphere, the temperature is approximately minus 25 degrees, the sort of heat to be found on top of Europe's Mont Blanc during wintertime.

The reason that temperature rises in the stratosphere is because ultraviolet light from the sun loses some of its energy to such molecules that exist at that distance from the Earth. The temperature therefore rises, but the region is hardly hot, as we understand the term, owing to the general lack of gas, or indeed of anything, at that height and temperature. There is ozone up there, notably between 10 and 20 miles, this form of oxygen having been formed by solar radiation. The created ozone decays into ordinary oxygen in time, but a big question (about which much more in Chapter 13) is the extent to which human activity is hastening this process of decay, thus causing less ozone up above and therefore more ultraviolet radiation to reach the ground. In doing so it also reaches us, with our skins sorely damaged by too much of this kind of light.

Climate, in its essence, is what happens on the other side of our front door. Whatever is happening in the troposphere and higher up may be relevant, but the events taking place within our immediate surroundings are the dominant concern. As with trees, which have to stand and experience whatever assaults them, we too suffer drought, rain, flood, heat, cold, fog, ice, breeze and gale whenever such conditions chance to come our way. Our advantage, as with burrowing animals or those using shelter of any kind, is that we too can escape the severest and most unwelcome climatic assaults.

Normality is what we all, whether burrowers, home makers or mere shelterers, have come to expect, and we can cope accordingly, but freak years are the most influential, often by being lethal. A year which, however transiently, is too extreme for a particular lifestyle can abolish that form of livelihood. Vegetation can speedily be killed, thus also killing all creatures dependent upon that vegetation, and therefore all other creatures dependent upon them. Human beings may survive a year or two without successful crops, and may recover from freak happenings, but it is the climatic peaks and troughs which, in general, do the most damage. Every now and then, and quite suddenly, a chain of events can combine to create a particular form of devastation. El Niño, about which also much more later, is the abrupt warming of a Pacific current with knock-on effects all over the globe. Climate is a mass of dominoes, each ready to fall in any of a bewilderment of directions when various forms of nudge set them toppling.

Nevertheless solar radiation is at the root of Earth's climate and our understanding of it. It might be expected that the quantity of sunshine received in each locality is entirely straightforward, being dependent upon latitude and the time of year. To a certain extent this is true. The actual equator receives its minimum quantities at midwinter and midsummer when the midday sun is precisely over one or other of the two tropics, Cancer or Capricorn. Similarly it receives its maximum amounts at the two equinoxes of March and September. As for the middle latitudes, known as the temperate world (of either north or south), they receive the least of their heat in their respective winters and most in their separate summers. The two poles receive absolutely none in each of their winters and most in their midsummers. So much for simplicity.

Now for confusion. When it is midsummer's day in the northern hemisphere the equator, as already stated, is then receiving less sunshine than at any other time of year except for midwinter's day. At the northern tropic – of Cancer, lying at 23.5 degrees N. – there is on midsummer's day the combination of a vertical sun at midday *and* a day length of 13.5 hours. The day is therefore hot, with a lot of solar radiation for a long time. North of that tropic the sun is never overhead, whatever the season and whatever the time of day, but the days are longer than those lasting only 13.5 hours at the tropic of 23.5 N. However the declining angle of the sun for each northwards degree is more than matched by the increasing length of day. This region north of the northern tropic therefore receives more sunshine than anywhere equatorial, a quantity which continues to rise until the latitude of 43.5 degrees N. This is about the level of Sarajevo, Florence, Nice, Marseilles, Toronto, Milwaukee, Sioux Falls, Idaho Falls, and much of southern Oregon.

Summertime day length is even longer further north of that Oregon-to-Florence latitude, but this increase cannot compensate sufficiently for the decreasing height of the sun. The quantity of sunshine therefore falls north of that latitude of 43.5 degrees N., despite the increasing lengths of daylight. It continues to fall until the latitude of 62 N., this being not far from St Petersburg, Stockholm, Bergen, Oslo, Helsinki, Anchorage and southern Greenland as well as the top of Hudson Bay.

Therefore, whatever one's previous opinions, and even firm convictions, all places on the latitudes of 43.5 N. and 62 N. receive the greatest and the smallest quantities of sunshine during midsummer's day in the northern hemisphere. (For convenience this generalisation assumes no cloud cover to amend its basic truth.) The same situation also occurs in the southern hemisphere, with 43.5 degrees South being the level of New Zealand's Christchurch, Tasmania's Hobart and Argentina's Chubut province. As for the southern 62 degree latitude, which experiences least sunshine on its midsummer's day, that line lies to the south of every permanently inhabited southern region and just grazes the northernmost areas of the solitary northward projection of Antarctica, the region known as its peninsula.

The increasing day length with increasing latitude, which reaches 24 hours at the Pole, does help to compensate for the declining midday height of the sun. Even at midnight on midsummer's day at the North Pole the sun is still 23 degrees above the horizon. This fact explains why sunburn is so easy in polar regions whatever precaution has been taken. Not only is ice or snow a good and searing reflector of sunshine, as on any ski slope, but humans in either arctic zone are remorselessly subjected to radiation whatever the time of day. There is no kind of let-up, as helpfully occurs when darkness falls in other regions where the sun does not shine for 24 hours a day.

To sum up. Solar radiation is more complex in its powers than might be expected. The thickness of the atmosphere through which it has to pass further disturbs simplicity. The troposphere in which we live is colder at altitude above the equator than above the poles. Earth's oceans and land-masses each behave quite differently, whether they are being heated by the sun or are cooling when it is absent. Every area of land surface behaves uniquely. The amount of water vapour present in air not only changes in its proportion but also alters the properties of that air. Every form of land obstruction, from a single bush to a mountain range, disturbs the flow of air. And all of this is happening on a gyrating planet which itself sets up eddies that behave contradictorily in either hemisphere. No wonder, to say it yet again, that the swallows are one week late, or London has three inches of rain in four hours, as has occurred, or that crops can fail abnormally and abysmally.

There is also the matter of microclimatology. Even if we lie down, as opposed to standing tall, we encounter different conditions, certainly less wind, probably more water vapour, and either more or less heat according to the nature of the land on which we lie or stand. This fact helped to initiate a cornerstone of meteorology, the Stevenson screen, designed by lighthouse engineer Thomas Stevenson, father of author Robert Louis Stevenson. This now familiar box-like object, always painted white, with twice-louvred sides and a double roof some five feet above open and preferably grassy ground, was initiated by the infant Royal Meteorological Society in 1866 as a standard measuring chamber for containing instruments, thus minimising as far as possible all local variation.

Despite the perturbations, the conflicting influences, the micro and macro zones, as well as the vagaries of coldest, hottest, windiest, driest, wettest and every other extreme, there are still some basic facts of planetary life. On Earth there are the polar regions, and the temperate areas, and the single tropical zone straddling the equator. These three need to be examined independently in order later to examine, and hopefully better to comprehend, the planet's changing circumstances. Its climate, which we tend to take for granted, has changed, is changing and will undoubtedly continue to change. No year is identical to its predecessor, and never has been. There are also cycles, of cooler times becoming warmer times, of wetter periods becoming drier ones, of ages which were icy and others which were hot.

The difficulty, of course, is to know what lies at the base of any alteration. Nothing happens by chance. There has to be cause. We human beings now have the arrogance, or even the humility, to believe that our myriad activities may be one kind of cause. The atmosphere is some ten million times heavier than we are, the six billion of us on Earth. It might therefore seem inconceivable that we could amend it in any fashion, like individuals attempting to quell or even hasten a hurricane, but there are facts we must confront. It is therefore better to meet them after taking note of what is, or should be, in order to judge more competently what is no longer true, and should not be.

Chapter 5

THE HUMAN YEAR

Human beings do not live everywhere, but make a good shot at doing so, being the most ubiquitous of all species. They thrive up to 14,000 feet. They live in heat and cold, wherever they can grow their crops or hunt their food. Around the Pacific are some extremely isolated islands with only a few dozen inhabitants. Even deserts, not only hot but also cold and generally inhospitable, are not immune. The only major piece of Earth without human settlement is the five million square miles of Antarctica. There are scientific bases, permitting overwintering, but no one truly lives on that coldest continent and no baby has ever been born there. The place was probably too far from other land for anyone to make the leap. South America's tip is nearest, being only four degrees of latitude north of Antarctica's projecting peninsula which reaches up to 63 degrees South, but Tierra del Fuego – of today's Chile and Argentina – is sufficiently challenging as an environment to inhibit thoughts of travelling even further south across a forbidding stormy strait, the stormiest in the world.

Charles Darwin, first to take proper and scientific note of the native Fuegians living at that tip, was amazed both by their hardiness – they wore hardly any clothes – and by their apparent reluctance to improve their lot beyond the barest minimum. He would not have credited them either with ambition or ability to adventure southwards, certainly not by sea. Besides, no true land mammal had made the journey to exist on that frozen land, no creature equivalent to the terrestrial species of bears and wolves existing in the north. No modern human adventurer even knew any territory was there until it was first seen in 1820, despite this tremendous continent forming 26% of all the land in the southern hemisphere.

The Arctic is quite a different situation. Land animals do live there –

polar bears, foxes, hares, caribou, wolves, musk-ox, lemmings. It is warmer. Slabs of the northern continents reach up to within seven degrees of the North Pole, whereas the southern tip of Tierra del Fuego is still 34 degrees from furthest south. The northernmost portions of Asia and America are not continuously connected with the great landmasses lying to the south of them, but fragments of these continents serve as stepping stones, aided by the frequently frozen channels of sea between them. The Arctic region is therefore less immediately forbidding as an area into which to trespass, albeit cautiously, island by island, bit by bit.

Nevertheless, although the Inuit have made this area their home, most of the Arctic is not what the rest of the world considers welcoming, being exceptionally cold and also dark for much of every year. (Eskimo is a derogatory term, used by nearby Indians to indicate raw-meat-eaters.) The Inuit domain is impossible for crops, and their hunted food is often in short supply. They did not have a script until missionaries created one during the 20th century, and it is difficult imagining even a literate igloo-bound individual keeping a journal of the annual events which he knows only too well; but, if such a diary had ever been kept, it would have outlined the seasonal Inuit calendar more or less as follows:

January – No sunshine, but a glow to the south at midday affirming the sun's existence elsewhere. Quite good weather, with either gales or calm. Very cold. By end of month can begin to see colour of dogs. Seals still catchable at their breathing holes. Female bears sleeping, *not* hibernating, but males still hunting. Birds very, very scarce.

February – More snow than in January, more gales, and some Föhn winds. These are warm, with heat caused by press of cold wind descending from the hills. The sudden melt often disastrous, as clothes outside igloo become wet before being frozen, and caribou encounter vegetation covered by hard ice rather than softer snow. Föhns often followed by intense cold. Days can be colder even than in January. Fogs common. Hunted animals extremely thin. Midday sun now visible.

March – Sun much higher every day. People getting about more. Bear footprints turned in, meaning skinny bears, rather than out. Their youngsters still in dens. Meat in general tastes bad, but fur is in best condition. Lemmings have first brood, and will repeat process monthly. Carnivores therefore grateful. Daylight plentiful, and even excessive – able to damage eyes – at month's end.

April – 'Spring' but no flower, no leaf, no insect yet seen here in the north. Birds beginning to appear, notably snow buntings. Fulmars arrive, and start choosing sites, but wait for May to begin nesting. Good gatherings of our neighbouring friends and relations. Caribou begin migration. Ground squirrel, Arctic's only hibernator, wakes up. Much mammalian mating. Bear cubs out, with mothers. Skins no longer satisfactory. Harp seals good quarry, with pups not swimming for first three weeks and easy to catch.

May – Now 24 hours of sunlight in north. Good moving time for us. Colours which were invisible in darkness are now difficult to detect in brightness. Midday temperatures often above freezing, but never at midnight. Seals even pant in heat, and can die from heat exhaustion. Greater warmth means unwelcome rain, often lengthily. Snow-storms also frequent. Musk-ox calves born. Moulting begins. Mammals ahead of birds in breeding, with early-bird snow buntings not yet laying eggs. Tremendous rush of migrants, but weather still cold and difficult with more birds dying during this month than any other. Plankton flourish. So fish busily feeding, and seals busy feeding on fish. Insects now plentiful. Seals and bears mate at this start of summer. Many flowers in the south. Traps useless as so much food around.

June – Flower month in north. Insects abundant, such as butterflies in south. Mosquitoes particularly noticeable, ready and willing to feed all day long. Mammals numerous. Egg-hatching a commonplace, with almost all eggs cracking before end of month, save for leisurely eiders still only laying. Previously empty cliffs now packed with birds. Atlantic whales arrive. Sun shines brightly for 24 hours, this being solstice month. Igloos no longer suitable, and people move into tents.

July – True summer. Life everywhere, with all links of food chain snapped into place. The feasting is universal, and numbers considerable, but fewer species (of everything) than further south. No reptiles live here, and only few amphibia by Arctic circle. Some birds, such as female phalaropes, already heading south after only six weeks in area, leaving young to fly south with males later. Most birds busy feeding young, assisted by abundance of food and considerable day length. Musk-ox hair moult everywhere. Sea ice still present with moving floes making navigation difficult. Glaciers speed up, and icebergs common.

August – Least sea-ice. So ships can travel furthest north. Almost all birds depart. Exceptions are fulmars, slow to start and slow to leave. Still many flowers, but general feeling that summer is almost over. Autumn is as short a season as spring had been. Nights shrinking rapidly. Only 81 days between 24 hours of daylight at solstice and 12 hours at equinox, or almost 9 minutes a day difference on average. But actually speeds up with approach of winter. Rain very frequent. All forms of moulting now finishing. Adult geese extremely vulnerable as unable to fly until moulting concluded.

September – Winter arrives before end of month. First frosts kill all flowers, with some dying colourfully. Insects and birds all gone, with earliest arrivals – the snow buntings – being last to leave. Streams and lakes freeze, but seas less so. Snow on high ground. No one yet in igloos, although southerners tend to stay in tents all year round. Caribou start to mate. Winter-white animals, the foxes, stoats, lemmings, hares, now change from previous greys and browns. Pelts become good again.

October – In northern areas the sun disappears before end of month, but it was equinox only one month earlier. Therefore day length is shrinking by almost half an hour per day. Hard to hunt with kayaks as there is so much ice, but ice not yet good enough for smooth sledge travel. Bears seek out dens, and squirrels make their holes before ground becomes too hard. Igloos made, and happily entered as warmer than any tent. Birds have arrived at their winter destinations. These greatly varied. Wheatears and snow buntings both nest on west of Greenland, but the buntings head for American Great Lakes while wheatears leave for Britain and Europe. East Greenland buntings all head for Europe or Russia. Every trapped ship will stay trapped for winter. Great sunsets even if sun no longer seen.

November – True winter again. The sun has gone, and snow and ice now rule. Good time for visiting neighbours, with stored food still abundant and hospitality easy. Snow is good for travelling, and real cold not yet begun. Aurora borealis often brilliant, creating more light and shadow even than the moon. Animals rarely seen, with lack of food for them now more devastating than lack of warmth. Dens well occupied, and animals start becoming leaner unless there is sudden abundance from, perhaps, a whale carcass.

December – Still not as cold as January and February will be. Darkest month, but moon and northern lights most helpful. Often calm conditions. Therefore still good for visiting, for inspecting traps, for bear hunting, and for harpooning seals at their blow-holes. Solstice is time of greatest darkness with the whole Arctic seemingly desolate, but it is also the time when rebirth begins, when the glow from sun in the south begins, but oh so gradually, to become a little brighter each day.

December is a time of nothing, but also an occasion when all the signs indicate that the annual cycle is about to start once more. In only 180 days there will be brilliant sunshine lasting all day long, the mosquitoes will be annoying, the sea-bird cliffs will have changed from emptiness to raucous

crowdedness, and the Arctic world will have witnessed a total transformation from the cold, silent and apparently lifeless place it is when the
12 so extremely different months are finally concluded.

Temperate dwellers to the south may wonder how human beings survive
such conditions, hearing of little which might induce them to mimic such
a lifestyle, but the Arctic year is only an exaggerated form of their own
annual cycle. December in the temperate world is also the time of greatest
darkness, of least growth, greatest hibernation, least reproduction, fewest
flowers, and an absence of numerous birds, so many having flown to
warmer regions every autumn. The Arctic difference lies in the absolute
alterations which occur each year. There is continuous daylight followed
by continuous darkness. There is an emptiness of life and then wholesale
abundance. There is feasting and then fasting, and a time to visit neighbours or do nothing of the sort. The Inuit are said to rejoice when the sun
reappears. They are also said to celebrate when the sun vanishes once
more, with freezing times more appealing than the halfway times of slush
and thaw.

There are some real advantages in living further north than the
temperate zones, such as fewer human rivals and much less likelihood of
conflict, let alone of war. The elements are undoubtedly forbidding and
can kill, but there is less disease, less infection, and less need for resistance
to combat virulence in general. Many an Inuit, if travelling south to
warmer areas, can promptly fall foul of the first disease to strike. The Arctic
is generally good for hunting, particularly during the crowded times when
all species are cramming a year's activity into the few active months. There
is summer and there is winter, with little intermediate. It is positive and it
is negative, with the seasons most precise. There can never be doubt in the
Inuit mind which time of year it is.

The Antarctic is similar astronomically, in that day lengths vary in
identical style according to latitude. The solstices are equally separated by
six months, but almost everything else is different, and how intriguing it
would be to transport some Inuit from north to south in order to discover
and observe their reactions to a wholly different but very similar place. The
title of Arctic, named for the Great Bear constellation, whose 'pointers'
currently aim for Polaris, the north star, gave rise to the Antarctic, the
opposite-to-the-Arctic.

And so it is, in almost every way. The place is land rather than water. It
is high rather than flat, even possessing an active volcano. About 90% of

the world's ice sits on Antarctica, with most of the rest on Greenland. There are 8,930 feet of ice beneath the South Pole surface, and only 15 feet or so of it at the North. Bergs 90 miles long have been recorded in the south, with their Arctic rivals formed from glaciers microscopic by comparison (even if able to sink RMS *Titanic*). The Antarctic is far windier than the Arctic, with one southern expedition recording an *average* wind speed of 50 miles an hour throughout the year, as well as 85 miles an hour blowing for 24 hours, including a gale averaging 107 mph for eight hours. Even 200 miles an hour has been recorded. That was no hurricane or typhoon, but the sudden descent of cold air from a higher plateau.

Antarctica's mammalian life, notably of seals, flourishes in its millions, of sufficient abundance and ease of capture to unhinge the mind of any transplanted Inuit. The southern continent possesses penguins around its boundaries in even greater quantities than the seals. Some 40 species of bird have been identified down south, but only 17 of these have ever nested on the Antarctic mainland. All of them, save for the sheathbill, have webbed feet, emphasising the sea's role as the crucial source of food. Botanically the southern continent is less exuberant than the north, mainly because so little of its land surface, perhaps only 1% even in midsummer – loses its covering of snow. There are no trees, or herbs, or woody shrubs, or ferns, and only two flowering plants, one a pink and the other a grass, but there are 70 species of moss and 350 of lichen. The largest animal of any kind to live entirely off the land is a fly less than three millimetres long.

The surrounding seas are explosive by comparison. There are six species of seal, seven of whale, a huge variety of fish, with one kind, the ice-fish, being the only vertebrate without red blood cells, and a tremendous mass of plankton generally known as krill. Euphausid shrimps, *Euphausia superba*, are its principal components. They do most to satiate mammalian hunger, despite being a mere 2.5 inches long and translucent, appearing as if short of nourishment for others. Adélies are the commonest Antarctic penguins, living in jostling, noisy, smelly and enchanting 'rookeries'. Emperor penguins endure the coldest conditions of any bird when breeding, with strong winds and freezing temperatures assaulting them after the females have laid their single eggs each southern autumn.

Efforts have been made to introduce penguins to northern latitudes, partly to serve as replacement for the Great Auk, first bird to be called a penguin but clubbed finally to extinction in the 1840s. (The etymology of penguin is obscure, perhaps being Breton for white head, perhaps after pin-wing, or after the Latin – *pinguis* – for fat.) These transplantations of the southern birds were all unsuccessful, with no one really knowing why the experiments failed. Penguin species as a group are not too choosy where they live, with the total range of this family Spheniscidae extending from the Antarctic to the equator, although most of the 16 penguin species are found between latitudes 45 and 60 South.

Despite this range of habitat they did not take to the north even though, over a century ago, there was much human encouragement for them to do so, a series of failures are considered good news by modern ecologists who resent any unnatural and man-made introductions. Even as late as 1936, when the individual concerned should have known better, a Norwegian introduced nine king penguins to northern Norway, with the apparent blessing of Norway's National Federation for the Protection of Nature. Fortunately – but not for the penguins – they did not do well. One moulting bird was killed by a woman who presumed it was ill. Another was caught by a fisherman on his hooks while a third, thought to be a troll, was promptly put to death. (So that is what trolls look like.) The final penguin, leaving no kin, is believed to have died in 1954. There has been recent alarmist talk that polar bears could be introduced to Antarctica, the better to control the huge colonies of birds and seals. Nothing so drastic has yet happened. Even huskies have recently been banned from the continent, a sad loss for anyone who has savoured husky-powered travel, but it was thought some might escape and wreak carnivorous havoc. (There is also the disadvantage that dogs need feeding every day whereas tractors consume fuel only when working.)

Fuegians living at South America's southern tip felt no inducement, most understandably, to travel further south. Even modern man was daunted by this least welcoming of the continents. Captain James Cook succeeded in reaching 71 degrees South, but failed to travel further owing to the 'press' of ice. He assessed correctly that ice would exist all the way to the South Pole, but never discovered any of the land lying beneath it, despite believing in its existence: '. . . yet I think there must be some (land) to the South behind this ice; but if there is, it can afford no better retreat for birds, or any other animals, than the ice itself'.

A portion of the continent's actual land was not observed for another 46 years, with a variety of claimants (American, British, Russian) seeing its peninsula in 1820. No one chose to overwinter anywhere in that for-bidding region for a further 78 years, these earliest of ship-borne pioneers achieving this distinction by mistake. When the trapped vessel was released during the following summer its Belgian crew hurried north at once, grateful for renewed liberty. Both Roald Amundsen and Robert Scott, together with their teams, overwintered from 1910–11 before their

separate forays to the South Pole, arriving there in 1911 and 1912
respectively, but no other human visitor encountered 90 degrees South
until an American plane was flown over it in 1929. No plane actually
landed there until 1956.

Both polar regions are particularly susceptible to global warming. Those
of us living elsewhere are less immediately aware that one summer is,
perhaps, a single degree warmer on average than its predecessor. This is
not so, either for the Inuit in the north or the scientists in the south. They
are all acutely aware of changing ice conditions, of differing berg
formation, of glacial retreat. A one degree rise in temperature is not
immediately catastrophic but is all too noticeable. Climate is everything in
such places, and the greatest shouts about global warming have come from
south and north. No doubt they will continue to do so, and temperate ears
should hear them. The two arctics are like outposts, first to see and to take
note of some climatic change. They are our most sensitive zones, like
antennae, like feelers probing ahead, and serving as crystal balls informing
us what we can expect.

An Inuit story from Baffin land, described in *The Arctic Year* by Freuchen
and Salomonsen, can tell of that knife-edge on which they already live.
Several residents had to make a journey even though no sledges were
available. They therefore compromised. Caribou skins were folded over
ice which had been carved into the shape of runners, and suitable cross-
bars were fashioned from frozen meat. Urine was then applied to the
runners for these to travel more easily. In short, ten out of ten, so far, for
adaptation. All went well until a Föhn wind arrived when they were
sleeping, this deluge of air from some higher plateau not intrinsically
warmer but with its extra pressure causing temperatures to rise. The dogs,
sleeping outside and near the makeshift sledges, took immediate advantage
of the softened cross-bars, causing a difficult situation to become critical,
even by Inuit standards.

With travel now impossible the dogs were eaten first, then some clothes,
and only two women were still alive when help arrived. There were
gnawed bones scattered on the ground for all to see. 'Have people been
eating people?' asked the rescuers, and received no answer. One woman
then ate voraciously of food brought by the newcomers, and she soon died.
The other was more cautious, and survived. Later she remarried and
produced four children, the number of her offspring among that original
sledging party. 'Now I don't owe anybody anything,' she announced after
the fourth was born. An Inuit debt had been paid.

*

The temperate year does have parallels with the polar year, in that there is more darkness in winter, most brightness in summer, and a greater warmth when the sun is more apparent; but there are not the absolutes either of total sunlessness or 24 hours of daylight. Nevertheless the temperature extremes can be considerable, with a greater range than is experienced within either of the arctic circles. The Antarctic becomes colder than any place on Earth, but warm days even on the northerly peninsula rarely rise above freezing. Winter temperatures in eastern Siberia can be almost as low as the lowest in Antarctica, but summer days often reach human blood heat of 37 degrees. Polar regions experience little precipitation, with six inches of rain equivalent being an Antarctic average, but that kind of quantity can fall within a single day in the so-named temperate world. Such places, having been deluged, can then experience zero rain for months on end. Dictionaries define temperate as 'moderate' or 'self-restrained', but the temperate world is often nothing of the sort.

Even so, and despite their frequently intemperate nature, the regions between the two polar zones and the broad tropical belt, which spans 47 degrees, are firmly called temperate. If this word is redefined as an abbrevation for temperature the title makes more sense, with its three sub-divisions each defined by warmth. The warm temperate regions, closest to the tropics, are often extremely hot but are more distinctive in never being cold. Their coldest month each year is never cooler than an average six degrees C. Therefore plant growth is possible throughout the 12-month period in a belt lying, roughly, between 30 and 40 degrees of latitude in either hemisphere.

This warmest of the three temperate regions can become extremely hot in summertime, notably inland rather than near the sea. Conversely it can become colder inland than near the sea, the presence of water modifying each possible extreme. Most amazingly, the daily range in temperature can be greater than the average yearly range, notably in the deserts. A clear sky plus a dry soil empty of vegetation will lead to great heat loss every night, and then to instant heating when the sun arrives once more. The so-called semi-deserts are often in warm temperate zones and are not particularly amenable to human habitation. (As the schoolboy allegedly wrote: 'The centre of Australia is so hot that its inhabitants have to live elsewhere.')

The year-long ability for plants to grow within the warm temperate world cannot occur unless there is suitable rainfall. In general those parts of this zone nearest to the equator have a more pronounced dry season, and therefore a longer pause in plant growth. Longer-rooted plants are most likely to survive, being able to tap moister soil during dry periods, and their leaves are often thick, waxy and curled, these all being contrivances to discourage transpiration and therefore water loss. Plants in cooler areas are usually soft to the touch, as well as unspiky and not unpleasant for humans wishing to lie down. Not so in warm temperate zones where spines, as around much of the Mediterranean, can forbid such intended relaxation. Try reposing on spinifex, the tussocky 'grass' covering swathes of central Australia and, as the schoolboy said, it is better to think of reposing somewhere else.

Much of the warm temperate world looks arid, and is arid by comparison with moister areas, but it is not so much the lack of rain which is paramount as those dry months when hardly any falls. Yearly precipitation in inches for Gibraltar, Athens, Palermo, San Francisco, Jerusalem and Alexandria is 35, 15, 25, 22, 24 and 8 respectively. Save for Alexandria these totals all compare agreeably with the 20 annual inches experienced by most of eastern England, but no one could ever confuse the soft, moist and gentle quality of East Anglia with those dry and dusty communities located in the warm temperate zone. They all have at least two consecutive months without more than a single inch of rain, namely 3, 2, 2, 4, 6 and 8 months, again respectively. It is those 8 dry months rather than its 8 annual inches of rain which make Alexandria such an arid place surrounded, save for the death throes of the River Nile, by desert land. Eastern England almost never experiences a month without an inch of rain. Any month-long rainlessness is even less probable during the warmer time of year.

Temperate region number two is the cool temperate. As this encompasses all of Britain, most of Europe, the northern half of the United States, northern China, Japan, southern Russia and the southern portion of Canada, where practically all Canadians have their homes, it embraces a huge number of individuals who consider their cool temperate region to be quite the most interesting climatic region of them all, so much so that other areas are somehow abnormal, freakish, and not strictly relevant to climatic argument.

By definition this cool, and well peopled, temperate zone has at least one month, but not more than five, whose average temperature is below six degrees C. Plant growth therefore occurs for 7 or, at most, 11 months in each year. This is also the zone where the four seasons of spring, summer, autumn and winter are at their most expressive. There is the gradual onset of growth as winter slowly yields, and there is a similarly leisurely retreat when summer's warmth and light are slowly replaced by yet another wintertime. Spring and autumn are therefore just as significant portions of each year as wintertime and summertime, a fact which properly applies neither to the polar regions nor to the warm temperate world.

Agriculture is very much a cyclical business in the cool temperate zone. It is also quite protracted. Crops are either growing, or about to be sown, or waiting for warmer weather. Harvest-time is not a single flurry, taking place within a single month, but occurs as summer begins to recede and as autumn starts to flourish, but all harvesting must be concluded before winter brings its chill. The coldest times become more pronounced the more each continental area is furthest from the sea, from Britain to Siberia, from Vancouver to Winnipeg, and Oregon to South Dakota. The internal regions *know* they will have winter. The coastal regions are not so sure, with the first cold snaps taking them almost by surprise.

This is partly because their winds, blowing either warm or cold, can come from any direction. The prevailing weather pattern is from west to east in this cool temperate area, and it embraces a considerable number of depressions, the cyclonic low pressure systems which line themselves up, for example in the Atlantic, to assault the land, such as western Europe, with awesome regularity. Although air gyrates around each depression the actual wind at any particular location, however much the cyclone generating them is travelling from west to east, can lead to a local wind from any other point of the compass. Hence the rapid changes in temperature, such a feature of the cool temperate zone. One moment the local wind is coming from the south. Then, as the cyclone moves eastwards, that wind can come first from the east and then from the north. The local residents are therefore mystified – what to wear, what to expect, whether the car will start. So too the forecasters who hedge their bets accordingly, saying it will be mainly dry, but perhaps wet, possibly cloudy with sunny intervals, and about average for the time of year. As for warmth and cold in general within the continental areas the two extremes are most likely to occur soon after the two solstices of June and December. In the coastal regions the times of greatest cold and heat are more likely to be August and February owing to maritime reluctance, first to warm and then to cool.

The cool temperate year, so gentle in its changes and so very changeable, does have attractive characteristics. Nowhere else, not even the other temperate regions, is springtime so pronounced. Neither is it so extended elsewhere. Snowdrops can appear in January, perhaps thrusting through snow to herald the start of another year. Daffodils and crocuses are later harbingers, pre-empting the true spring months of March, April and May. Many a plant, reacting enthusiastically to any warmth in that season, can then be savagely punished by late frosts, these perhaps choosing May to strike. Certain trees, as if with greater wisdom having seen it all before, are not fully in leaf until May is out, until certain that the vicissitudes of spring are truly past. So when does spring occur? During the six months from January to June is one answer. For many years such an elastic definition can seem most reasonable.

By that sixth month everything has started. Bird courtship possibly began even in winter. The hibernators warm up, probably in March. Hares leap about madly that same month in sexual rivalry. The first migrants arrive, just in time to see the first emigrants depart for further north. By then most of the resident birds have already started nesting. Amphibians lay their eggs in March and April. Fish start to spawn. Insects begin to be abundant, and reptiles bask in any available sunshine. There is a new effulgence, a new flowering, but not so explosively as in the Arctic. It happens, but takes its time. Most importantly, unlike the Arctic, there is time to spare. Consequently, as if obeying Parkinson's famous law, work expands to fill the time available for its completion. The cool temperate spring, being such a protracted business, fulfils that rule precisely.

The young of many mammals are born in May or June. By then birds are either with a nest, with eggs, or with young. The earlier birds may later produce another clutch, or even three in all. These earliest of the summer months witness most egg-laying by fish. Amphibians may continue their egg-laying until August. Insects are extremely numerous, in good time to be consumed by all the insectivores. Flowers blossom in keeping with other events, perhaps to flourish before leaf canopies cut off most light, or to take advantage of insect pollinators, or before rivals have consumed the available space. There is no law demanding when flowers should bloom, and cunning gardeners can maintain colour almost throughout the year.

June sees the longest days of all, with most days warm and anti-cyclonic. Depressions can still occur, still moving from west to east and bringing winds from every direction, but these breezes are gentler and the depressions less deep. Warmer days mean more clouds and more rain than in wintertime, but fewer wet days. August floods can be particularly devastating. The range of a cool temperate summer, perhaps with fruit blossom killed by frost, with ripe corn too wet to cut, with soggy vegetables spoiled by rain, and late potatoes ruined by cold, can make one wonder over its cool temperate merits, but it does grow grass superbly. Nowhere else seems as capable of making good grass and gracious lawns as the temperate British Isles.

September, with the first signs of autumn, is often drier than August. The migrants tend to leave, but their loss is softened by other birds returning, these having nested further north. There is still abundance, of caterpillars, wasps, flies, spiders, as if summer is unending. Weeds love this time. Mushrooms and other fungi can also do extremely well. October can seem better even than September, with an 'Indian' summer often occurring long after every Inuit knows absolutely that winter has returned. Cool temperate leaves can turn slowly brown, as if wishing to drag out autumn. Only those, as in New England or Canada which are suddenly subjected to wintry chill, will produce the wonder of a glorious fall. Depressions get deeper, bringing wind either from a cold Arctic or the warm south, thus confusing humans, animals and plants whether summer or winter is now in charge.

There is still uncertainty in November, with the days short and about to be still shorter, and the month neither particularly warm nor particularly cold. Even December, home to the longest nights of all, can have days surprisingly warm as well as extremely cold. Animals are not fooled by such contrariness, with hibernators firmly hibernating, and insects either dead, dormant or as eggs. Resident birds continue as before, suffering whatever weather comes their way. Large mammals do not store food, save as body fat, but smaller ones may do so, their personal reserves too inadequate. Fish swim deeper but all water-living creatures are better able to survive owing to an odd property of water. It does not continue to contract in volume when cooled below four degrees C. but actually expands. With warmer water then being heavier it sinks to create a zone of greater warmth beneath the ice. The frozen water up above serves as protection and insulation for the warmer water down below. Snow also acts as insulation, perhaps covering up the food supply but discouraging further heat loss from the ground.

Air temperature does not have to be below freezing for snow to reach the land. It may even be as high as six degrees C. if conditions are right, with unsaturated and windy air, but a day of three degrees will generally see some large flakes if, of course, precipitation is occurring. The lower the temperature the finer the snow. Britain's geographical outline only ranges from 50 North to 59 degrees, but is more than sufficient to show correlation between latitude and snowfall. Cornwall has 5 days on average on which snow is seen, the Midlands 15 days, Edinburgh 20 days, and the Scottish Highlands 25–30 days. On far fewer days, to the disgust of children but relief of almost everyone else, does snow actually settle.

Height is as relevant to snow-cover as it is to plant growth. In Scotland there is, again on average as with so many of these statements, an extra day of snow-cover per year for every 24-ft rise in altitude. Altitude can be seen as a corollary to latitude. Scotland above 2,000 feet has a covering of snow equivalent to the low Alpine resorts ten latitude degrees further south. On Scotland's higher reaches snow may remain for 100 days or so. British mountains would be permanently capped with snow if Ben Nevis were above 5,300 feet, if the Lake District exceeded 5,900 feet, and North Wales 6,300 feet, instead of its actual 3,560 feet. Had such a calculation been made before Mt Kilimanjaro was discovered for Europeans by European man there would have been less disbelief that its 19,340-ft peak possessed permanent snow, despite being only three degrees north of the equator.

The British obsession with weather, with Britons steadfastly telling each other what anyone with even a single sense organ would already know, does have cause. With snow in April, frosts often in May, cold nights in mid-summer, most rain in summer, hot days in October, and even balmy ones in December and January, it might appear that the four seasons are a myth. It *can* be colder in June than in December. Snow *can* crush daffodils more than the much earlier snowdrops, and Indian summers *can* be succeeded, six months later, by entire crops of apple blossom frozen into nothingness. British weather *is* worthy of comment almost every day of the year.

Charles II said the English summer was three fine days followed by a thunderstorm. Were he alive today he could still, on occasion, find his words most apt, with summers equally uninspiring, often most unsummery and never behaving as they should. No wonder, therefore, that Britons and others, equally upset by climates which so often go awry, are ready to welcome the notion that weather is now different, that the huge long-lasting snowmen we used to make are no longer possible, that the days of all our childhoods are now firmly ended, and that floods, droughts, gales and storms have increasingly arrived as varied substitutes. What is going on, we ask, as 200-year-old oaks are suddenly uprooted? Why does a tornado hit Selsey Bill in Britain's normally soothing south? And what about those clifftops, and even well-mapped villages, being eaten by the sea? 'It isn't right,' say citizens as they wade to their front door, or even wonder where their front door has gone. In every age and in every place, or so it is easy to suspect, there must have been mutterings that exceptional times were in command, but never have the mutterings been so loud, or so well vindicated, as those being uttered now.

*

The third temperate region, known as cold temperate as against warm or cool, has at least six months colder on average than the critical six degrees C., that powerful borderline between plant growth and the lack of it. There are, as a result, no deciduous trees in this area, the business of producing leaves and then shedding them being unrewarding with such a limited span of warmth. Instead this zone's trees are conifers, the ever-greens which in general keep their needle-leaves from one year to the next. Much of Siberia and of Canada is cold temperate, without either the benefit of a lengthy summer or the biological outburst of polar exuberance when a year's activity is densely packed, so far as this is possible, into the warmest months.

The 62 degree latitude, already mentioned for its minimum insolation, runs through the cold temperate zone. The combination of a low sun as well as the shorter daylight hours of summer than occur further north in the northern hemisphere causes this region to be less welcoming than either the warmth of further south or the abrupt vitality of the polar regions. Few people live in the planet's cold temperate areas which embrace most of Scandinavia as well as the emptier and colder portions of Siberia and Canada. All the major cities of Canada exist within 150 miles of the United States, despite this second largest country of them all stretching from 42 degrees N. up to 83 degrees, a vertical distance of 2,830 miles. As for Siberia, which a friend described as 'three games of chess in a jet', its size is bewildering, covering as it does one-twelfth of the Earth's land surface.

A lack of enthusiasm for living in the cold temperate zone is wholly understandable. (So too its ready-made suitability as a punishment zone, notably in Russia.) The darkness of winter is one reason, with daytime then such a brief event. The cold is another, with Canadian motorists plugging in their cars at parking meters to prevent them freezing during a shopping trip. Even then, and after starting, there will be a steady drubbing from the tyres until the portion which had been flatter during the waiting time becomes rounded once again. Truck drivers, knowing of the difficulty in starting frozen engines, can keep their motors running until, say, April. Canadians and Siberians do work throughout their northern winters when mining, drilling, or even logging, but tend to live in surroundings, often underground, as unlike the real and outside cold temperate world as can possibly be contrived.

Human beings are not only the most ubiquitous of all species, but also the most adaptable, whether using modern methods or making do, as with the Inuit and tribes like the Chukchi in north-east Siberia, whose skills, with cunning and courage, permit them to survive. Survival is such a problem there is little opportunity for any other concern. One reason for conceit about the importance of warmer temperate areas is that so many of the great civilisations have flourished within their boundaries. The most advanced communities, as one climatologist stated, 'have evolved in lands where the year goes round with a strong, but not excessive, seasonal rhythm punctuated by considerable weather changes'. The cold temperate excesses can be more than sufficient impediment to advance in general. India, home to many civilisations, is exceptional, with half of this sub-continent being tropical and the northern half warm temperate, but its particular form of seasonal punctuation is the presence or absence of rain rather than the existence or lack of warmth.

Whereas the various temperate distinctions are somewhat blurred, despite those precise classifications of warmer or cooler months, the tropics are clearly defined by the canting of Earth's axis. Within the tropical zone, stretching from 23.5 degrees South to 23.5 degrees North, the sun is definitely overhead at midday at some point during each year. The tropical belt is therefore similar to the two arctic regions, in that all three have an astronomical basis to their definitions. The arctics experience permanent daylight for at least one day per year whereas the tropical day is never longer than 13.5 hours and never shorter than 10.5.

Nevertheless there are some other climatic definitions for the zone between Cancer and Capricorn. A hot and tropical place, as distinct from warm temperate, is said either to have an average annual temperature of at least 21 degrees C. or no month with an average temperature below 18 degrees. Frost can occur within the tropics, and often does in deserts, but daytime heat – perhaps searing up to 50 degrees – more than compensates to give a high annual warmth. For plants the growth temperature of six degrees is maintained, and growth is therefore possible provided the cold snaps do not kill.

In fact the world's hot areas, thus defined, are slightly greater than the actual tropics as defined by latitude. In broad terms these warmest places exist on the equator's side of the high pressure barrier lying between 30 and 40 degrees, either north or south. The trade winds flourishing on each side of the equator blow from east to west. At about 30 degrees from the equator in either hemisphere the winds are not only gentler but more variable, while on the poleward side of this less decisive region are the 'westerlies' which blow from west to east. In general, therefore, the hot tropical regions experience not only a near-vertical sun at midday but winds travelling from east to west.

To an Inuit the dominant seasonal event each year is the presence or absence of sunshine. To temperate dwellers it is the existence or lack of warmth. To tropical inhabitants it is the occurrence or absence of rain. The word monsoon even means season, this being *the* annual event. Some 60% of the world's people rely on that regularity to provide the water which they need. Precipitation is always significant but not every tropical region experiences rain in similar style, let alone similar quantities. In the doldrum regions, either around the equator or at the interface between westerlies and easterlies, some rain tends to fall throughout the year. Between those doldrum areas is the tropical division, so-called, which experiences considerable rain only in the hot season, save for this division's marine areas where rain is more consistent. There is also the desert division which is similar seasonally, save for having hardly any rain.

Tropical rain is not easy to delineate, either month by month or from one year to the next. (Once, when wishing to motorcycle through Africa without benefit of rainstorm, of mud-deep highways or mere flood, I examined charts to time my journey. My hope was to travel through each region on the drier side of sodden. Then I read: 'It should be remembered that the African climate is in general uncertain, and wet weather may be experienced during the periods shown as dry and vice-versa.' I promptly folded the charts before setting off in a more relaxed frame of mind, aware that planning would probably be ineffectual. In fact I was to experience rain only once during those four months of two-wheeled travel, that rain falling during the peak of an area's dry season.)

Equatorially there are, or tend to be, rainy maxima in April and November, with one bigger than the other. These usually occur shortly after the sun has been directly overhead. Elsewhere within the tropics are variations on this general theme, such as the smaller/lesser rains being non-existent or, furthest from the equator, both sets of rains coalescing to form a single rainy deluge. Temperate rain, as temperate dwellers know full well, can fall at any hour of day or night, with no particular preference. Tropical rain can be pleasingly predictable on a daily basis, even if the wetness is occurring during the dry time of a year. The day may start bright, with clouds soon forming until they have swollen sufficiently to drop their rain, perhaps at three pm or thereabouts *every* afternoon. The sky then clears for a dry evening, with humans happy at the ordered sequence and able to plan accordingly.

It so happens that the world's two largest river systems not only straddle the equator, but are large partly because they do so. The Zaire, or Congo, river receives its heaviest rain in the north during May, and its heaviest southerly rain during December. Therefore the main river is well supplied all year round, and does not experience the ignominy, perhaps preventing navigation, of a dry season. The Amazon is similar, its main river course almost coincident with the equator and therefore receiving water from north and south during their separate rainiest times. This river's catchment area is even bigger than the Zaire's, and experiences 3,000 cubic miles of rain each year. Almost everything about the Amazon is difficult to absorb, its eventual discharge to the Atlantic of 11 million cubic feet per second being 11 times that of the Mississippi. Many of its tributaries are over 2,000 miles long, and those 3,000 cubic miles of water per year should be compared with the 36 or so cubic miles received annually either by Pennsylvania or England & Wales.

The well-watered tropical regions are therefore very suitable for plant growth and for animal life exploiting all that vegetation. They are warm all year round, and never suffer lengthy periods of darkness. The lack of a pronounced season means that plants, notably the trees, are not uniform either in shedding their leaves or in flowering and fruiting. Almost any stage in these cycles can be observed at any time of year. Near-permanent growth means that tree-rings, so noticeable in sawn timber from temperate regions, are scarcely visible. Tropical forests are not dominated by a few species, as with the oak or beech woods of Europe. Instead it is possible to find as many plant species within a few square miles as exist in all of Britain. Such diversity is not readily understood, as speciation is particularly encouraged by less optimum circumstances, but it certainly exists. Finding one kind of tree is followed by considerable effort to find another of the same species. They are hardly ever side by side.

Agriculture is a very different business than further north or south. There is not a growing season followed by a period of zero-growth, but an ebullience which makes crop-planting much more casual. Chop down a bit of forest. Put a few yams or cassava among the fallen debris, collect their produce a short while later, and move on when the soil seems drained of useful nutrients. The drier time of year is an occasion for greater harvesting, but nothing like so fixedly as in the temperate world. One tropical disadvantage is that a field, however unworthy of the name for Americans or Europeans, can become a thicket of new growth if left untended for a while.

Unlike the polar darknesses, which are entirely predictable, and the temperate coolnesses, which are also to be expected even if not quite so cool or warm as the previous year, the arrival of tropical rain or its non-arrival is much more of a lottery. Unreliability increases according to distance from the equator. Calcutta's rain at 22 degrees N. varies annually by 16%. Lahore's rain at 32 degrees N. is not only much less but varies by 38%. On the Indian subcontinent crops are planted in expectation of normal rain, even though abnormality is quite normal.

With so many people living in risky situations the possibility of disaster is the only certainty. A yearly precipitation of 30 inches, an average for London or around North America's Great Lakes, is marginal in the tropics, partly because evaporation is so high with most rain arriving during the heat of summer. One-tenth of an inch in Britain, the average for those days on which rainfall is recorded, is scarcely sufficient to lay the dust in hot areas, being undetectable one hour after its arrival. If landing on hot rocks or bare ground it can vanish almost instantly. Or, if the rain is sudden and the ground too hard for seepage, the much needed libation can speedily reach the streams and rivers without any benefit to the land on which it fell. There is one advantage in the fact that rain falls during the heat of summertime. Ambient temperatures are definitely cooled by the arrival of so much moisture, with May in general being a hotter month than either June, July or August which are the rainy months. When the monsoon finally comes to an end the wintry months have arrived once more, thus preventing temperatures from rising above 24 degrees or so.

Tropical areas other than the Indian subcontinent also experience unreliable rainy seasons, partnered by equally unreliable dry seasons, but the slab of land projecting into the Indian Ocean merits most of the superlatives. As described in *Climatology*, by A. Austin Miller, the Indian subcontinent lies 'between the greatest landmass and the warmest sea, shut in by the loftiest mountains, (and) backed by the highest plateaux of the world'. Its situation therefore 'provides the best conditions for a monsoon'. In winter the whole region is a high-pressure zone, with hot dry air being gently expelled, but summer's arrival dramatically switches everything. The high-pressure centres north of the Himalayas, part of Asia's tremendous continental system, abruptly change to become a major low. Air has to rush in from somewhere to fill this emptiness, and it arrives from India's south-west, this air having been travelling over some 4,000 miles of warm water. It is therefore saturated, and brings not just relief but a deluge. Average arrival date at Bombay is 2 June, in the central provinces 10 June, at Bengal 15 June, in the eastern provinces 20 June, and the end of the month for Delhi. For the previous seven months Bombay has received a pathetic average of 1.3 inches. It is then subjected, again on average, to 20 inches in June, 24 inches in July, 14.5 inches in August, 10.6 inches in September, and 2 in October. This means that a rainfall of 70 inches in five months becomes less than 70 millimetres in the seven remaining months. If this tap is turned on too late, or is turned off too soon, many crops, such as rice, will fail. (How odd that 'It never rains but it pours' is an expression arising in, and often used in, England of all places.)

Bombay is at the coast and low-lying. When the monsoon's hot, wet air arrives from the south-west it travels inland. There it is forced to ascend, and a lowering of its pressure makes it become even more generous with its rain. One hundred miles to the south-east of Bombay, where land is 4,500 ft above sea level, this hillier country receives 2.5 inches of rain every rainy day. In that upland area, and during the first four months after the monsoon has begun, there are 116 rainy days out of 122. During one third of the year, the wet third, this region receives 250 inches, this being the only portion of the year with any rain worth mentioning. As Bombay receives a mere 70 inches annually, and the nearby hills receive 3.5 times as much, that indicates an extra inch of rain falling every year for each 25-ft rise in altitude. One zone may therefore be marginal for a particular crop. A touch higher up the slope, or a little lower down, and that crop may no longer be possible, the rainfall either excessive or quite inadequate.

The switch from high pressure north of the Himalayas to low pressure, causing the south-west monsoon winds to blow with their annual generosity, must be the result of a multitude of influences, each adding up to that momentous alteration. Earth's orbit is at the root, with the midday sun right over the equator during the mid-June summer solstice, but countless other adjustments must be at work, hastening or slowing the monsoon, and helping to bring abundant life or some form of destitution to the thirsty areas.

When there is talk of global warming, with its concern about mankind's adjustment to the natural situation, it is easy to be additionally concerned about the on-off nature of the monsoon. It already seems such a chancy business, coming soon or late, and profligate or miserly. So what will human interference do to this risky annual rhythm? Will the rain be lessened or increased? Will its timing be hastened or retarded? And will the monsoon's form of bounty become even less predictable? The livelihoods of hundreds of millions of people exist in that marginal world, with even India's population now in excess of one billion. Any change is almost bound to be deleterious, with so much hanging so precariously on that chancy and annual event known as the monsoon. It is *the* critical event, each and every year. Nothing will change that, but almost any change will create more apprehension and, quite possibly, an even greater quantity of misery.

*

A desert is the converse of a monsoon area. Monsoon regions expect rain, and suffer tremendously when this is inadequate. Deserts expect nothing of the sort. They are normally dry but, on occasion and unexpectedly, they do experience rain. Those who live in deserts assume nothing save unrelenting sun, and are probably delighted when some rain does fall. It is therefore arguable that desert life is less hazardous. There is no optimism, no faith, no expectation, and therefore no devastating loss as can occur in areas with monsoons. Desert living is always difficult, save when very slightly easier. It is simpler to assume the worst. (A dictionary's definition of desert is most straightforward: 'deserted; desolate; uninhabited; uncultivated.').

Most of the world's deserts are not on the equator, despite that belt's proximity to the sun, but exist between 15 and 40 degrees of latitude in either hemisphere. Most such barrennesses are in the north because most land is in the north, with the Sahara by far the greatest at 2,600,000 square miles. Second largest are the 1,100,000 square miles within Australia. Most of the other large patches of arid land – Turkestan (900,000), Arabia (480,000), Colorado (195,000), Gobi (180,000) and Thar (74,000) – are in the northern hemisphere. Exceptions are Argentina (400,000), Kalahari (90,000) and Chile (74,000). Of Earth's 58,000,000 square miles of land a total of 7,250,000 are classified as desert, with a further 2,300,000 as semi-desert. (Some intra-galactic traveller, reporting back about Earth's terrestrial potentialities, could well be unenthusiastic about our planet's formidable quantities of ocean, of ice, of tundra, of rock – and of unrewarding desert.)

A desert's definition has exercised many minds. For most of us a desert is a place that looks like one, being hot in daytime, barren of vegetation, and either sandy or with such dry and dusty earth that it might as well be sand. The essentials are heat, leading to considerable evaporation, and modest rainfall, with the combination of heat and rain at the base of almost all the definitions. For example, a place is said to be a desert if its average heat is 21 degrees, or more, and its rainfall is less than 14 inches. It is also a desert, by another yardstick, if the rainfall, measured in centimetres, is a lesser figure than the average yearly temperature measured in degrees Centigrade plus 16.5. The British average temperature is 10 degrees and its rainfall, in the drier areas, is 50 centimetres. Therefore, by that second definition, it is not a desert because 50 is a larger figure than 10 + 16.5. The rainfall would have to be less than 26 centimetres/10 inches for it to qualify.

Evaporation is extremely relevant to deserts. A bucket left in an English garden, or indeed anywhere in the cool or cold temperate zones, is likely to have some water in it after a spell of time, although less likely in summer than in winter, despite the greater summer rainfall. A bucket in a desert will probably be bone-dry very shortly after a shower has done its best to moisten the neighbourhood. In central Algeria the ratio of evaporation to rainfall is about 60 to one. Therefore rainfall would have to be some 60 times greater for there to be any hope of finding water within a bucket. In the nearby Libyan desert, where rain fails to fall during most years, the annual evaporation rate is 158 inches. Assuming no rain, which is the most realistic assumption, a bucket or tank full of water at the start of any year would have to be over 13 feet tall for there to be any water at the year's end. Those who built the Aswan High Dam in southern Egypt caused the River Nile above it to become a 200-mile lake. They also caused phenomenal water loss with at least 13 feet of that huge surface vanishing every year, an impressive five cubic miles or so.

The desert environment is undoubtedly harsh, but satisfactory for those plants and animals with suitable adaptations. In fact it possesses about as many species of birds and mammals, and even more reptiles, than exist in the nearby moister climates. Human beings, said to be a tropical species, have never been called desert animals. They are so profligate by comparison with true desert creatures, notably in their use of water. The desert jerboa never drinks and lives on grain and seeds. Humans not only perspire avidly, but lose a pint each day merely in their breathing and another pint in urine. A man doing no work but sitting in the heat of 50 degrees C. when humidity is 30% needs a pint of water an hour to maintain his perspiration, and therefore sufficient for him to keep alive. If the air is dry, and there is a slight breeze in such a temperature, he will need much more. Heatstroke, and the inability to keep bodily warmth at a reasonable level below 40 degrees, will kill him if sufficient liquid is not available to drink and then perspire. The jerboa is not only nocturnal, emerging from its burrow in the cool of night, but has a low metabolism, which means less heat production internally. It also produces dry faeces as well as concentrated urine. Even more to the point, when heat is unavoidable, the jerboa's own temperature can rise to 45 degrees before death will intervene.

Plant species are not as numerous as in moister regions, but deserts can suddenly be awash with flowers. Seeds, bulbs and tubers can survive dry months, or years, but will leap into life should a shower dampen their immediate vicinity. Then, when drought returns once more, they will slump back into a desiccated form which can endure. Other plants, less ephemeral in their survival technique, exist with roots of tremendous length able to tap low-lying water. Their leaves are either small or non-existent, and there is chlorophyll even in their twigs. The famous camel-thorn is a good survivor in desert conditions, save when that other great survivor, the camel, comes across it.

The desert season may not have the regularity either of the polar night versus the polar day, or of the temperate warmth versus cold, or even of the monsoon rain versus drought, but such arid areas are certainly hotter in summer than in winter. They are richer with life only when rain has chosen to fall. This refreshment may not be annual, or even most frequent during a particular month, but it is seasonal in that there are dry times and wet times, however much these may be independent of the yearly cycle.

Once again, as with any alteration to the planet's climate, the deserts are likely to be affected. Change is less likely to be catastrophic because so few people rely upon deserts for their livelihood. The greater, and more alarming, change would occur if the deserts were greatly to expand their already considerable areas. The dry lands around them are so marginal, with their people living so close to the brink, that any extra dryness would push them over the edge of feasible existence almost instantly. An inch less rain in England would perhaps cause adjustments to be made, and for some grumbling to become a touch more shrill. A similar lack around the deserts could cause a desperate and meagre form of livelihood to become impossible.

An impossible situation can occur at any extreme of normal circum-stances. Similar kinds of extremity can occur in all terrestrial zones, when rain either fails completely or is horrendously excessive, when either heat or cold is overwhelming, or when any aspect of weather is being outstandingly vicious and demanding. There are also exceptions even to the exceptions, when something so awesome happens that it merits another name. The term normally applied to any such natural event is disaster, the topic now ahead.

Chapter 6

NATURAL MISHAP

Expectation – predictability – major horrors – insurance crash – the storm called Mitch – unnatural effects of natural mayhem

'We can never have enough of nature,' wrote the early conservationist, Henry David Thoreau, no doubt on some balmy evening in the 1850s while a gentle breeze was ruffling his hair and nearby leaves. Millions would disagree with him, such as all who have suffered 200 mile-an-hour hurricanes. Or watched orange-sized hailstones fall upon their crops and homes. Or seen whole communities drowned by mud-slides, by avalanches, or in floods. Or watched as houses, cars and people have been wrenched skywards by tornadoes. Or witnessed volcanoes, tsunamis, earthquakes, firestorms and other forms of havoc destroying the landscape. On such occasions it is surely very easy to have had enough of nature, when nature is showing an obverse face to its gentle ruffling of hair and leaves.

We earthlings expect, in general, this year's weather to be much like the last and it is, usually, more or less the same. We expect reasonableness, that there should be some wind, and perhaps a gale or two, but not tree-uprooting, roof-discarding, car-transporting, human-slaughtering and devastating storm. There should be wetter times and drier times, but not the deluges which turn rivers wild, and then all nearby solid land into lethal liquid lake. Nor, conversely, should there be a total absence of water, resulting in dead crops, dead cattle and dead people with extreme simplicity. As for the planet as a whole, the orbiting, revolving, cooling and cracking thing on which we live, we all know it is still unstable, but are aghast when earthquakes release pent-up energy, when volcanoes pour forth some stuff of which our Earth is made, and when tidal waves cross oceans to cause mayhem, perhaps 10,000 miles from their starting point. Even whole islands can explode, as Krakatau did on 27 August 1883, or can suddenly arise, as did

Surtsey when first seen above sea level off Iceland on 13 November 1963.

Such happenings can come as bolts from the blue, or so we say, but there is always physical cause for their madness. The warm Atlantic does spawn eight hurricanes a year, on average. These do arise in late summer and autumn, when the sea is warmest, and they do revolve anticlockwise while travelling in a westerly or north-westerly direction. As for the Earth cracking along its various seams and fault lines, these lead to a million Earth tremors each year, but are most destructive only in regions where, in general, they have already been damaging.

There is a measure of predictability about all such catastrophes, these occurrences which suddenly assault a few or tens of thousands of us. Tornadoes do not happen everywhere, most of them, and certainly the most damaging, forming within the United States. They tend to occur in the middle south of that country from April to June, notably during afternoons and evenings. Contrary to hurricanes they usually travel from south-west to north-east. As with youngsters driving cars and then having accidents, their mishaps do apparently come from the blue and are often seen as extreme forms of bad luck; but, overall and year after year, youngsters do have more of them. The climatic catastrophes are the greatest extremities of weather, and can seem like accidents, but they follow general patterns much as the more ordinary forms of weather adhere to basic rules.

It so happens, along with fears of global warming, which is assuredly proceeding, that the quantity of major disasters, which seems to be increasing, may also be a consequence of human mismanagement. The year 1998 was labelled as the world's worst-ever year for natural catastrophe, notably by the insurance companies, its 12 months being three times more damaging than the average years of three decades earlier. In 1998 alone there was Hurricane Mitch, which killed 11,000 and caused five billion dollars-worth of damage, mostly uninsured, and there were also:

fires in Greece ($675 million),
fires and heat in the US (130 deaths, and $4 billion),
heat, up to 49.8 C. in India (2,600 deaths, 'or perhaps twice as many'),
floods in China (3,600 deaths, $30 billion),
mudslides in Italy (150 deaths),
floods in Bangladesh (4,500 deaths, $5 billion, with a formidable half of the country underwater).

Most losses are uninsured, occurring as they do in the poorer regions of the world, but the insurance bill of 1998 still reached $15 billion. As for deaths, the 700 'large-loss' happenings of that single year killed 50,000 people.

Britain still remembers, with understandable horror, the night of 31 January 1953 when the neighbouring North Sea rose eight feet higher than normal to surge over much of eastern England, thereby drowning 307 people, largely in Lincolnshire, East Anglia and Canvey Island. Holland also remembers that night, and with even greater horror, as countless numbers of its precious dykes were breached causing over 2,000 citizens to die. It was the eighth such surge to have occurred in the North Sea during the first 53 years of the 20th century, and the most destructive such event since 1897. Nothing so lethal of a natural kind has happened either to England or to Holland in the half-century since that dreadful surge, but the numbers it killed and the destruction it caused were modest compared with countless tragedies at other times elsewhere.

Galveston, Texas, suffered 6,000 deaths when a hurricane struck in 1900 after another 6,000 had already been killed elsewhere in the Caribbean. In August 1992 Hurricane Andrew caused an insured loss in Florida and Louisiana of $16.5 billion, and total damages of more than $30 billion. Some 300,000 individuals died from a typhoon's wave in India in 1737. An even greater number was killed in Bangladesh from a similar storm in 1970. The 'great hurricane' of 1780 devastated the Lesser Antilles and 22,000 died. Some 25,000 were killed when a Colombian volcano erupted in 1985, one of the 600 volcanoes around the world with the ability to strike. Messina in Sicily was virtually destroyed in 1908 when 77,000 people perished. Even during the recent century a total of 1.5 million died in earthquakes. The International Federation of the Red Cross has estimated that the number of people to be 'affected' by natural disasters will climb from 129 million a year to 220 million a year by 2006.

Such colossal figures can numb, particularly when involving other regions. I remember the scene in a UK reporter's room when someone queried, oh so casually: 'That Indian flood yesterday, was it five hundred dead or five thousand?' Loss of life, and loss of property, make initial head-lines, but 'loss of future' should be added. Mitch's 11,000 deaths did not include the 'thousands', perhaps reaching a similar total, who were reported missing. Nor all those who will quietly die, later on, from home-lessness and lack of shelter. 'Millions' were involved in that tragedy, suffering either instantly, or from drinking polluted water, or from subse-quent poverty with their fields and crops destroyed, or from escalating disease, or from a terrible cocktail of them all. Insurance can seem of minor

relevance when thousands of dead are being reported, but uninsured livelihoods and vanishing incomes are as relevant as lives. Loss of cash and
absence of crops kill no less assuredly. They merely take a little longer.

Hurricane Andrew not only murdered and destroyed, but it depleted
much of the available insurance capital for catastrophes. If a large cyclone
or earthquake is to hit a major city in these modern times the insurance
losses could easily reach $100 billion, which is more than all the
reinsurance capital. Property values are rising, and property itself is
proliferating – along with people. Solely within three counties of south-
eastern Florida the population has increased sixfold in 40 years, from
733,000 in 1950 to 4.4 million in 1990. Therefore the damaging 1926
hurricane which cost $105 million would now cause an insurance bill of
$72 billion. The current total pool of catastrophe capital is a reported $6–8
billion. For the general reinsurance industry it is about $30 billion, and
about $250 billion for the $40,000 billion insured by primary companies.
One devastated city – Miami, for example – could therefore also devastate
much of the world's economy.

If terrestrial warming *is* leading to more catastrophe, thus causing bigger
claims than the $16.5 *billion* paid out for Andrew whose strongest winds
missed Miami, some form of global upset becomes more probable. As with
the financial events of 1929 and that famous Wall Street crash, such
market failures affect far more than financiers. The whole world is
implicated, with the poor, in general, becoming poorer and therefore
death a frequent consequence. A hurricane does kill outright. It can also
kill from its reverberations as these travel round the globe.

Tropical storms seem to occur in cyclical fashion. There are some 80–90
around the world each year, with about half reaching hurricane or cyclone
strength involving sustained winds of 33 metres per second/74 miles per
hour, and with most of them occurring in the North Pacific. Quite a few
do not reach land, and only 10%, or fewer, form in the North Atlantic, but
it is these storms off eastern North America which have been most actively
studied (and most publicised), with reliable records of their savagery going
back to 1900. Miami was founded in 1896, and the city experienced a
trouble-free first three decades. Then came 17 hurricanes in that area
between 1926 and 1965, with five of them severe. A further period of
relative calm then followed before Andrew stormed ashore in 1992,
informing the younger citizens of Florida what their parents were talking
about when remembering hurricanes. Andrew's fury seemed to initiate a
further bout of cyclonic activity because there were 17 in 1995 alone, the
highest number for any single year since the 21 of 1933. After 1995 there

has been little let-up. In 1998 there were 13 tropical storms in the North Atlantic, of which nine formed hurricanes. (Perhaps we should be grateful we do not live on Mars. In April 1999 the Hubble space telescope spotted a hurricane near that planet's north pole which, 900 miles long and 1,100 miles wide, was probably composed of clouds of ice.)

Here on Earth it is not crucial for a storm to be labelled hurricane, and be given a name, to create chaos. The disastrous Mitch, which so pulverised Nicaragua, El Salvador and Guatemala in 1998, started as a hurricane, and was therefore named, but was downgraded and reclassified as a tropical storm before ever encountering land. It had first formed on 21 October as the fourth deepest depression, at 905 millibars, ever to be recorded in the North Atlantic. (The hurricane record stands at 888, as against 870 for a typhoon in 1975). To begin with Mitch reached category five on the Saffir/Simpson scale, the biggest possible classification, but by 29 October its depression had filled to become 994 millibars. Its winds had therefore slackened to 60 miles an hour before they reached the Honduran capital. No longer was Mitch a hurricane.

That gale's catalogue of destruction only occurred when its moisture-laden air started ascending Central America's higher ground. It then dumped 24 inches/610 mms of rain in a single day on steep land stripped of much of its former vegetation and therefore no longer capable of absorbing such a vigorous libation. Rivers of earth and mud then slid down the mountainsides, encountering and transporting wretched shacks which should never have existed in such hazardous areas. Mitch's deadly mixture was therefore a combination of strong wind, of that gale's slow passage over the ground, of its well-saturated air, of the higher altitudes it encountered, of the torrential rain it dropped, and of the countless thousands of people down below who had been forced by unending poverty both to damage the land around them and then to set up home in quite the wrong locations. These people promptly suffered from the terrible amalgam.

The rest of the world may soon forget Mitch, now that it has left the headlines, but it was the most deadly storm to strike the western hemisphere in two centuries. Its rainfall was considerable, but not exceptional for hurricanes or other storms in the Atlantic basin. The three extra reasons for its lethal qualities, according to a report in *Nature* in May 1999, were that:

it arrived at the end of the rainy season when the land was already
    saturated,

pressure on land had caused many hillsides to be stripped of trees,
before it made landfall, Mitch had been predicted to move
northwards.

The local population was therefore unprepared either for its arrival or its
style of devastation. Further and future storms will usurp the front pages,
and possibly create even more superlatives describing yet more mayhem,
but Central America will not forget Mitch, not for years and years and
years.

The construction of homes on wrong and likely-to-be-lethal sites may be
caused by a lack of alternatives, or avaricious builders, or thoughtlessness
by all concerned, but much of the modern upset attributed to 'natural
catastrophe' is often the result of any single one of such human errors, or
all three of them. Natural disasters *are* more probable in certain areas.
Hurricanes do *hardly ever* happen, as Professor Higgins pointed out, in
Herefordshire, Hampshire and Hertfordshire, but they *do* happen elsewhere, quite frequently, and better precautions should therefore be
considered concerning all forms of possible catastrophe. People of
California, whose homes are built near fault lines, or within tinder-dry
brush country, or near steep-sided terrain, should not be totally amazed,
and then affronted, before becoming immediately desperate for insurance
plus governmental aid, if quakes tear their homes apart, or fires consume
them, or mud-slides hurry downhill through and over them, burying such
wrongly sited dwellings with effortless ease.

All disasters, however natural, have unnatural effects when so much has
been put in their way which is so very ripe for destruction. Catastrophes
are on the increase, yell the victims, and somebody should do something.
Perhaps Article 229 of the Babylonian building code, drawn up in 1700
BC, might be resurrected: 'If a builder builds a house for a man and does
not make its construction firm, and the house he has built collapses and
causes the death of the owner of the house – that builder will be put to
death.' Such a law would revolutionise the construction industry. Affluent
homes, whose owners and builders should know better, would not be sited
in unsuitable locations. The owners of poor homes would be advised to
build in more appropriate regions, just as they are currently advised about
hygiene, smaller families and contraception. Then, perhaps, there would
be less unnatural destruction by way of natural catastrophe.

However, back in the real world, the insurance nightmare might, if it
came to pass, be more damaging, and more speedy, than the effects of
global warming. An earthquake first rocking and then igniting, via severed

gas pipes, a rich community, or a tsunami drowning a prestigious waterfront in similar style, or a hurricane tearing the heart out of Miami rather than the shanty shacks of impoverished settlements, could be instantly devastating.

In theory the wealthier conurbations are better prepared, their structures sounder, and the early warnings of impending trouble more likely to be obeyed, but in practice even well-formed theories can be faulty. Earthquake-proof structures can, and do, collapse. There can also be ill fortune. Hurricanes can alter course perversely, causing some evacuations to be unnecessary and others to be too late. Avalanches can fall where they have never done before. Moribund volcanoes can suddenly, and terrifyingly, come back to life again. All such catastrophes tend not to happen in gentle or piecemeal fashion; that is not their way.

Global warming on the other hand will be more insidious, gradually changing ocean levels and steadily damaging the old stability, whereas bolts from the blue, or the sea or the ground, are as good as instantaneous. The world's economy could score a devastating ten on its own form of Richter scale. If ever a Hurricane Apocalypse tears into some famous and truly wealthy city it would not only be the most costly there has ever been, but could make financial institutions topple as easily as the structures housing them.

Disaster, in general, does not strike in areas where there is most wealth. There have been exceptions, as with Kobe in 1995, Messina in 1908 and, of course, San Francisco in 1906, this earthquake being Richter 7.9 (some say 8.3) and killing 3,000 individuals. Hurricane Betsy, which came ashore in 1965 to travel from southern Florida to Louisiana, caused $1.5 billion in damage while doing so. So how much financial damage would a Betsy do today, or another San Francisco, when there is more insurance than ever before, more humanity living nearby, more housing, more of everything? It would cause lots and lots and lots.

Now to some specifics about the various forms of nature, and of devastating happenings, which Thoreau disregarded when welcoming more of nature. Despite the plethora of information about to be listed, with winds, storms, vortices, thunderbolts and so forth doing their damnedest, these destructive elements are in every case a rarity. For more than 99% of days in any one locality Thoreau was quite right. Nature is not always gentle and beneficent, particularly when producing unduly heavy rain, prolonged drought, strong winds or heat waves, but the nightmare situations are relatively uncommon. If atmospheric physics were wholly different, and if Earth was quite a different kind of place, it could be that tornadoes were daily perturbations, that lethal lightning would strike again and again much like rain, and that winds could *average* 50 miles an hour as they can blow in Antarctica, but nature's extremes are rare, and rarely experienced by any one of us, wherever we may be. For which a very great deal of thanks.

Chapter 7

CATASTROPHE

Hurricanes and depressions — floods — tornadoes — 3 May 1999 — waterspouts — dust devils — hail — avalanches — lightning — Ben Franklin — volcanoes — dust veils — Krakatau and Tambora — earthquakes — MMS and Richter — Kanto and Kobe — tsunamis — wildfires — natural extremes

The several forms of extreme climatic severity can all be formidable, in terms of damage done and people killed. There is no particular relationship between their styles of disaster and chaos; nor is there any logic in the listed order which now follows.

Hurricane Originally a Carib word, then adapted by the Spanish before being transmuted into English, hurricanes are also known as typhoons in the western Pacific after the Cantonese tai-fung, or as cyclones around India, as baguios by the Philippines, and as willy-willies in Australia. To qualify, by whatever name, as the severest form of cyclonic storm they must have wind speeds of at least 74 mph/64 knots or force 12 on the Beaufort scale. Beaufort Force 13 is up to 92 mph, 14 up to 103 mph, 15 up to 114, 16 up to 125, and 17 up to 136 mph. There is also the Saffir-Simpson scale of five categories which has been used by the US National Hurricane Center since the 1970s.

1	more than 980 millibars	74–95 mph winds	Minimal damage
2	965-980	96–110 mph	Moderate
3	945-964	111–130 mph	Extensive
4	920-944	131–155 mph	Extreme
5	less than 920	more than 155 mph	Catastrophic

These categories are therefore graded in descending order of barometric pressure, as well as ascending order of wind speed, with the lowest pressures being the most severe.

As millibars are thousandths of one atmosphere, or bar, it might be supposed that average pressure at sea level would be defined as 1,000 millibars. In fact, notably in the world of aviation, a ground-level pressure of 1,013.25 mb (the customary abbreviation) is generally taken as standard. The unit 'hectopascale', a more recent arrival, is also favoured by some, with the conversion factor – say the flippant, although entirely accurately – between millibars and hectopascales being 1 to 1. As further definition 1 millibar is equal to 1,000 dynes per square centimetre, or 0.75 millimetres (0.02953 inches) of mercury. As for the standard atmosphere, that equals 760 millimetres (29.9213 inches) of mercury or that difficult number of 1,013.25 millibars/hectopascales. An atmosphere also equals 1.03323 kilograms per square centimetre or 14.696 pounds per square inch. A metric atmosphere (to round off this indigestible paragraph) is one kilogram per square centimetre which equals 735 millimetres of mercury – 28.94 inches, or 14.22 pounds per square inch.

Hurricanes are severe low-pressure areas originating over warm (27–8 degrees C.) oceans, with the resultant winds flowing in a circular fashion towards the centre. The lowest pressures involved, down to 900 millibars or so, represent a couple or more barometric inches, and therefore 150 lbs of load off every square foot of land surface. It is therefore small wonder that such tremendous winds are involved. The whole system can contain hundreds of thunderstorms, each gyrating around an 'eye' of extremely little wind, no rain and clear skies. This quiet zone is some 14–20 miles in diameter.

Hurricane storms arise in late summer and autumn when sea-water is warmest. The season is from June to November north of the equator – mainly August to October, and from December to May south of it – mainly February to April. The maximum wind strength may be 250 miles an hour, which can only be exceeded within tornadoes. Hurricanes embrace far greater areas, perhaps a million square miles, and can last much longer (three weeks). They occur in general between 5 and 30 latitude degrees on either side of the equator, but mainly between 10 and 15. They must be distanced from the equator for the Coriolis force caused by Earth's rotation to have effect, and this sets them spinning. Their turbulence can reach up to 50,000 feet in the atmosphere.

Each year some 80–90 tropical storms form above warm oceanic waters. Slightly more than half of these develop into hurricanes, with winds sufficiently fast and strong to qualify. Most originate in the Pacific – it is after all the biggest ocean of them all – and about 12% start in the North Atlantic, of which half a dozen or so become hurricanes. Many of those formed in the Pacific do not reach land but can certainly damage the shipping in their path.

For the creation of hurricanes a great mass of air must have high humidity and high temperature as well as variable lapse rates – the rate at which temperature falls with increasing altitude. Hurricanes in either hemisphere tend to track from east to west, before moving more to the north in the northern hemisphere, and to the south in the south. Climatic conditions over land, as against warm water, are no longer suitable for their maintenance, the supply of moist air having been cut off. On reaching land they soon diminish into tropical storms, and then into standard depressions. Even Britain, placed so far from the ravages affecting the south of North America, can be affected by the death throes of hurricanes when they have diminished into no more than ordinary depressions arriving from the Atlantic.

Much was made in Britain during October 1987 of a television meteorologist going out of his way to deny rumours that a hurricane was imminent. 'Earlier today a woman rang the BBC and said she had heard there was a hurricane on its way. Well, if you are watching, don't worry, there isn't.' Some 12 hours later, during the early hours of Friday 16 October, south-east England suffered its worst storm in memory, with insurance claims exceeding £1.2 billion. The death toll of 20 killed was low, mainly because the storm arrived when most people were securely in their homes. Some 15 million trees were blown over, many of them more than a century old and seemingly invincible. The storm's maximum recorded gust was 115 miles an hour at Shoreham in Sussex. Extreme depressions are much commoner in Scotland. Fraserburgh, for example, recorded a gust of 140 miles an hour only two years later, but damage in that area was minimal by comparison.

Perhaps it was the fact of various stock markets, including London's, stumbling near that time which made many residents believe old days had gone for good, particularly after a further storm hit southern England early in 1990. This second one took 47 British lives, and 50 more on the European mainland, plus £2 billion in British insurance claims. Trees were less hard hit on this subsequent occasion, partly because the rampaging winds covered a wider area and also because, being wintertime, the trees were no longer in leaf and soil was nothing like so damp.

In truth, although scant comfort to those affected, these were *not* 'warm-cored hurricanes', as correctly defined, but ferocious 'temperate-latitude depressions'. The English forests, rooftops, electric pylons, etc. did not know that, and fell over as if bona fide hurricanes had been in charge. The

1987 storm was probably the greatest since a similar depression had damaged England in November 1703. Daniel Defoe, later to detail Robinson Crusoe's near-solitary adventures, stated that his pen could not 'describe that great storm, nor tongue express it, nor thought conceive it unless by one in the extremity of it'. It was certainly severe, with 8,000 people allegedly dying as a consequence of its fury.

England's 1987 storm was subsequently eclipsed in its ferocity and damage by the gale which ripped through much of France on 26 December 1999. Some 360 million trees were knocked over at the root or, yet more disturbingly for any form of resale, snapped off centrally. Foresters who have studied the effects of both storms say that France's was 20 times more severe than England's. It certainly made the price of timber slump to one-fifth of its previous levels, causing immediate distress to forest-owners, sawmills and wood merchants. France, in general, possesses a greater covering of trees than England, but the damage done was all the more remarkable for occurring in midwinter when every deciduous tree had lost its leaves. The *tempête* also managed to kill 100 individuals, and was reckoned – with reason – to have been the most disastrous event in France since World War Two.

Hurricanes did not always receive personal names, with the lethal storm that hit Galveston in 1900 given no such identity. Female names were later allocated in alphabetical order, with the first storm of each season being an A and so on. Hence no hurricanes such as Zena, Zuleika or Zoe as 26 sufficiently violent storms do not occur in the North Atlantic during a single year. The custom of giving only girls' names was altered in 1979 when masculine names were also introduced, various pressure groups having had their say.

The 20th century's names were not particularly ethnic; so further adjustment was extremely likely. In fact, no sooner was this sentence written – in November 1999 – than it was announced from Washington DC – in February 2000 – that tropical cyclones in the Western North Pacific and South China Sea were to be named 'with familiar local words'. The director of the National Weather Service Pacific Region stated that 'using local names that people recognize should encourage them to pay more attention to the warnings'. The first storm of 2000 would be called Damrey, meaning elephant in Cambodian. What this word means, if anything, for other people with other Pacific languages is perhaps irrelevant, but does indicate a step in the right direction away from Henry, Alan, Peter, Jane and Co.

The gentleness of some of the earlier titles is at variance with the power and savagery which they identify. Andrew (a long way short of Elephant) killed 61 people in Florida and Louisiana, rendered 160,000 homeless, and left a 1992 damage bill of $30 billion, of which $16.5 billion was insured. Agnes, although only a category one hurricane, managed to kill 122 in 1972. Hazel travelled north in 1954, swamped Toronto and killed 81, thus creating one of Canada's very worst natural disasters. Camille, possibly the gentlest name of all, but a category five storm, took 256 lives in 1969. (It is the only category five storm ever to have struck the US mainland.) Betsy's damage in 1965 totalled $1.5 billion. Iniki was probably the best photographed storm when its 225 mile-an-hour winds came ashore in Hawaii during September 1992 to kill 2, injure 98 and devastate prosperous – and camcorder-rich – Kauai, the 'garden island' which is westernmost of the Hawaiian group. Ten thousand houses were then destroyed, one-third of the total, at a cost of $2 billion. (No one is yet suggesting fiercer names, with the deepest storms labelled Pizarro, Adolf, Bonaparte, or Genghis Khan, but many previous titles have seemed odd, and out of keeping, with the havoc they involved.)

Hurricanes, by whatever name, do their damage partly via the fearsome force of their winds but mainly, and in general, by their precipitation. The rain they drop, perhaps a couple of feet in 24 hours, can cause flooding, or mud-slides and erosion plus a restructuring of the land. Mitch, as already indicated, was not even wind force 12 when assaulting Central America; its rain caused all the hurt. The water dropped by hurricanes kills either by submerging the land, and the houses and the people, or from sudden torrents when the moisture so recently arrived chooses to hurl itself downhill, along with everything and everyone in its path. When India's low-lying state of Orissa was hit by a savage typhoon in November 1999 it was flooding that did the damage, first by drowning people, then by destroying crops, and finally by spreading disease, such as gastroenteritis. One way or another that storm killed 'thousands', caused two million to lose their homes, and ten million lives to be in jeopardy when desperate people drank water speckled with carcasses. The wind itself was relatively unimportant, save for driving so much water on to the land and for destroying fishing boats out at sea.

An odd fact, discovered by a group at Arizona State University after it had examined 50 years of hurricane data, is that storm wind speed is weaker at weekends by as much as 11 miles an hour. This coincides with an earlier revelation that 22% more rain falls near the east coast of the US on Saturdays than on Mondays. Levels of carbon dioxide and ozone both rise as weekends approach, and these are partnered by increased pollution which is thought to 'seed' offshore cloud formation. This could be the cause of greater rainfall, and might also remove some water vapour from the severest storms, thus diminishing their power. Therefore weekends do have different weather, often for the worse, a fact the rest of us have always known.

Floods are natural, along with flood plains, but the current human propensity to build homes near rivers, to constrict river flow between banks and dykes, to build dams, and chop down trees causing speedier run-off from the land, can aggravate the age-old natural tendency to flood. Greater damage and greater loss of life become inevitable when rivers no longer have open spaces to receive excessive water. Each natural flooding can therefore have unnatural consequences, and these may all get worse. Any warming of the planet, whether man-induced or part of some terrestrial cycle, will exacerbate the trend by melting quantities of ice, by therefore raising water levels, and thus making yet more land liable to flood.

Once again, as with hurricanes, there are formidable lists of death and damage. The Yellow River is said to have killed one million in 1938 after some dykes had been destroyed. The citizens of low-lying Bangladesh always suffer tremendously when cyclones come their way, such as the two million who died in 1970, the 2,000 in 1988, the 125,000 in 1991, and the 30 million who were rendered homeless in 1998 – with such well-rounded totals serving as powerful indicators that the precise tallies are quite unknown. In 1973 the Mississippi's banks failed following excessive rain, promptly turning much good land into lake, killing 11 people, and creating a damage bill of $1.2 billion. Twenty years later, as if unsatisfied with such destruction, the Mississippi and Missouri rivers inundated large portions of nine US states, killing over four times as many individuals and increasing the damage cost eight-fold. In Europe the River Oder burst its banks in 1997, drowning great portions of Poland and Germany, killing more than 100, and damaging three billion dollars-worth of property. Spring snow-melt that same year in Canada was similarly destructive when the Red River overflowed to flood major portions of Canada and the US.

So what can be done? Nothing (yet) can prevent or even hinder such low-pressure storms, but early warning of their formation and their likely passage can certainly help. The hurricane to hit Galveston, Texas, in 1900 killed 6–7,000 in that community, roughly 20% of the total population.

This death toll created the worst natural disaster in the nation's history, and has still not been exceeded within the United States, despite the huge population increases in similarly vulnerable seaside areas. Galveston's principal error lay in its being built largely on a sand-bar. Not only were its homes extremely susceptible to the 120 mile-an-hour winds coming straight from the sea, but some early flooding from the same storm prevented many of Galveston's inhabitants from leaving their island for the mainland when such an evacuation seemed more than sensible. In one community of 20,000 not a single house remained, and almost half of all Galveston's homes were swept away.

Even in poor regions, where it may be difficult moving people from threatened areas, much human life can be saved by preparation. Ten thousand died in Andhra Pradesh when a cyclone struck in 1976, causing those in charge to plan accordingly. Fewer than one thousand died in 1991 when a similar storm arrived. The evacuation of 690,000 people within 40 hours, a formidable undertaking, proved inordinately worthwhile with, statistically, one life saved for every 75 people moved.

Good forecasting of possible tragedy can therefore be extremely beneficial. Mitch killed 'thousands' in Nicaragua when an old crater was suddenly filled with water, an event that was quite unexpected. Earlier anticipation of likely possibilities via computer analysis would have cost very little in money, but saved enormously in human suffering. It will be 20 years before Mitch's devastated areas are even back at their former impoverished state. The Central America Disaster Appeal, when advertising for funds one year afterwards, stated: 'Hurricane Mitch roared through in 52 hours. Months later, it is still destroying lives.' The cost of redevelopment will dwarf any sum which might, or should, have been spent in anticipation, in predicting what a storm can do, in announcing the most basic precautions when a storm is imminent – 'Get out of the valleys, now' – and in educating people as well as institutions about possible outcomes should misfortune choose to head their way.

Or, for some, when good fortune arrives in similar style. The so-called divine or god-winds, which scattered the Mongol invasion fleets sent by Kublai Khan, saved Japan from their attacks in 1274 and again in 1281. These storms were almost certainly typhoons, and their intervention – turning probable defeat into undoubted victory – gave rise to kamikaze, this name meaning divine wind. Cyclones and typhoons can also be visually arresting for all who are able, and sufficiently secure in their location, to enjoy this luxury. In 1942 one observer in Bengal watched the sea retreat for 12 miles, a consequence of the pressure drop offshore. The sea then returned as a wall of water 30 feet high. However spectacular the sight – and it must have been fascinating witnessing a great quantity of sea-bed suddenly exposed – any such observer should not delay and marvel, but hasten at once to higher ground, 100% aware of the retribution about to arrive. This marine return will assuredly cover up, and most speedily, not only the exposed sea-bed but much of the local land as well.

Are hurricanes increasing and are they getting worse? It might seem so, judging by their activities in the early 1990s and yet more recently. The Americas were battered by 14 major storms in 1998, and a similar number occurred in 1999, with Floyd – forecast as 'the worst for many years' – in September of that year, although softening its rage before reaching land and doing real hurt. The US National Oceanic and Atmospheric Administration, NOAA, concluded that the North Atlantic hurricane season of 1999 was the fiercest on record, with five storms reaching Category 4 (when pressure had dropped below 944 mb and winds had exceeded 131 miles an hour). Fortunately three blew out to sea and two, including Floyd, weakened before reaching land. These numbers are twice the long-term average, and their increasing incidence could possibly result, in part, from human activity, along with global warming. Or, as cycles do occur, together with other trends and whims of nature, they may be happening more frequently in an entirely natural fashion.

Long-term evidence, with information collected from 1944 when hurricanes were first tracked, examined and given names, indicates an overall decline, in both numbers and severity, during that well-studied period of more than half a century. As with so much of the changing climatic story the jury is still out. Many scientists believe that current hurricane numbers lie within statistical norms, with the intensity of the 1990s balanced by a lull in the 1970s and 1980s. On the other hand warmer oceans are more likely to generate more hurricanes, with warmth being crucial for their creation. A computer at NOAA has calculated that an oceanic temperature rise in the western Pacific of 2.2 degrees C. will lead to wind speeds 12% higher. It would therefore seem possible, if not probable, that warmer air (caused by increasing greenhouse gases) will lead to warmer oceans and therefore to more aggressive and perhaps more numerous hurricanes.

Those who observe storms forming, such as the Hurricane Research Division of NOAA, consider that a hurricane's path is easier to forecast than its intensity, but hurricane Opal, of 3–4 October 1995, abruptly gave its scientists disturbing food for thought. The storm's vortex had been circling within the Gulf of Mexico for some days, but it then encountered a 'deep pool of warm water' which had become detached from the main

Gulf Stream current. Within 22 hours the leisurely storm then moved from Saffir-Simpson 2 to S-S 4. This transformation left the local administrators with insufficient time to organise evacuation. Besides, night was falling, and administration became more difficult, when the storm was gathering much of its new-found power. It seemed, to the watchers, as if Camille of 1969 might be repeated, the only S-S 5 ever to hit America. Then, providing much relief all round, Opal changed back again to an S-S 2 before encountering land at Pensacola Beach, Florida. It killed nine, and caused $3 billion in property damage, having already killed 50 in Mexico and Guatemala in its earlier stages; but it was no Camille.

The change in Opal's intensity, both up and down, had been impossible to predict, but the actual site of this storm's landfall was less of a surprise. On the other hand Hurricane Floyd in 1999, having promised to be one of the most damaging of all time, proved to be nothing of the sort. It certainly sent two million people inland, with 'Avoid Floyd' stickers on their cars, but it failed to come ashore anywhere in Florida, as initially threatened, or at Charleston, South Carolina, as was a later possibility. It eventually arrived even further north, having lost much of its venom during the landfall postponement. There is a danger that 'Cry Wolf' warnings, coupled with misjudgement, may cause people to be less enthusiastic about recommended evacuation on subsequent occasions. One trouble with direction forecasts is that a trifling shift of five degrees between suspected course and eventual track can cause all the difference. As a consequence there is either mayhem, death and destruction or, as with Floyd, infinitely less than had been feared.

Only one thing is for sure. Even if hurricanes blow no harder than before, and bring no more rain, they will become more costly. The most severe hurricane of 1926, which caused damage costing $105 million, would now – if equally strong and attacking a similar region of southern Florida – create an insurance bill almost 700 times as great.

The tornado, an intense revolving storm, can occur when a mass of cold air meets an opposing mass of warm moist air. A violently spinning vortex reaches down from the base of a storm cloud, notably when there are several thunderstorms in the area. Such a column is not a grown-up dust devil, those whirling dervishes of sand so often seen cavorting over heated ground. Instead the true tornadoes, named from a blending of Spanish words for thunderstorm and turning, only result from collision between major masses of warm and cold air, and they only merit the actual name tornado when their swirling funnels reach the ground.

As with the Richter scale (for earthquakes) and the Beaufort scale (for wind strength) there is the Fujita scale (for tornadoes). F0 involves wind speeds of 40–73 mph; F1 of 74–112 mph; F2 of 113–157 mph; F3 of 158–206 mph; F4 of 207–260 mph; and F5 of more than 260 mph. Theodore Fujita, of the University of Chicago, named the damage from these six types as Light, Moderate, Considerable, Severe, Devastating, and Incredible – with that last assessment appearing merrily unscientific. There is also the Torro scale which ranges more conventionally from one to ten. As tornadoes often destroy the instruments deployed to measure their activity it is therefore necessary to estimate the generated wind speeds, with their circular winds often substantially more than 200 mph and their updraughts 180 mph. Such viciousness makes hurricanes seem almost placid by comparison, with 'extreme' Category four hurricanes generating wind only up to 155 mph.

In general the width of a tornado storm is less than 500 yards and they last, also in general, about 15 minutes. Most are in the F0 category, existing for seconds rather than minutes, but a few are entirely different, such as a 1917 tornado which travelled 293 miles in 7.4 hours. A tornado's speed over the ground is usually about 35 mph, but can be much faster, as with the terrific twister of 18 March 1925 which covered 219 miles from Missouri to Indiana in 3.5 hours (over 60 mph) and killed 695 people when doing so, including 234 in the single community of Murphysboro, Illinois. On average tornado storms lead to 70–80 American deaths each year, despite warning systems and a proportion of people in vulnerable areas now possessing basement shelters within their homes. The trails of destruction left behind are, on average, about two miles long with a width nearer 1,000 feet. A mere 240 acres might be nothing compared to the considerable region a hurricane can lay waste, but the tornado's destruction is often tremendous and all-consuming with, perhaps, not a house left standing in its chosen path.

Tornadoes tend, as with hurricanes and fierce cumulonimbus storms, to happen in the summertime – March to July or even to October in the northern hemisphere, and vice versa in the south – but they can be contrary. One occurred in January 1969 when it killed 28 in central Mississippi. Traditionally they are afternoon or evening events and, unlike hurricanes in the north, they usually travel from south-west to north-east. They can come singly or in groups. The greatest spawning of modern times occurred on 3/4 April 1974 when a total of 148 tornadoes hit 13 US states as well as parts of Canada. The death toll from this most unwelcome two-day epidemic was 316.

Tornadoes only happen over land and are therefore more frequent in the northern hemisphere. The United States is particularly suitable for their creation, with warm wet air blowing from the Gulf of Mexico meeting cold air descending from the north-west. About 1,000 tornadoes a year 'touch down' somewhere in the US, when their vortex visibly links the ground with a cloud-base up above. The so-called 'tornado alley', a belt from Texas to Nebraska, witnesses most tornadoes and also most deaths. The state of Kansas is hit more frequently than anywhere else in the world and receives, on average, 24 twisting visits a year. Oklahoma is another state much favoured by this savage face of nature.

On 3 May 1999 a particularly vicious tornado group of some 50 elements, mainly classed as F4 but containing an extraordinarily fierce F5 component with speeds reaching a record – it is believed – 318 miles an hour, killed 47 people and caused $500 million worth of damage. This was despite good early warning, which could do nothing about the property damage but undoubtedly saved lives. Its most vicious swathe started at noon 70 miles to the south-west of Oklahoma City. Chickasha at 4.47 pm was an early victim when the tornado went straight through its municipal airport. Then came Bridge Creek at 6.35 pm with, even before this time, very good warnings being given, these greatly assisted by courageous ground crews closely, and sometimes very closely, following the destruction, their on-board radar measuring the twister's internal velocities. Thirty-minute warnings were given for those in the tornado's forecast path to stay indoors, get into the storm cellar if there was one, or into the bathroom if there was not, and certainly away from glass. Despite the general area being famous for tornadoes, most houses in that region did not possess such a cellar. The likelihood of any particular house being hit, with tornado swathes almost always less than one mile wide, is still small, and cellars, of course, cost money.

At 7.20 pm the F5 hit the edge of Oklahoma City at Moore. This was unfortunate in that these suburbs are thickly populated, but oddly fortunate with that city being the centre of so much tornado study. At one time even the forecasters were preparing to vacate their offices, with all non-essential personnel already despatched to safety. By then the trail of severe damage was 19 miles long, and Oklahoma City itself suffered the destruction of '8,000 homes, some 20–40,000 cars, 3 churches and 2 schools' but 'only 22 dead, thanks to the early forecasts'. The tornado then continued north-eastwards, with warnings reaching ahead to Sapulpa and Tulsa, but in the early hours of 4 May the twisting form of tumult 'suddenly disappeared', according to those still in their chase vehicles.

Scientists stated afterwards that they had learned much from the episode of 3 May, but not how to conquer or diminish an actual tornado. The only hope for the future is an even better and more accurate lead time for their forecasting, which still means having 'people on the ground'. The Oklahoma nightmare had lasted for over 12 hours, and was followed by an astonishing assortment of storm stories, of a baby being blown a hundred yards and surviving, of 4 by 2 inch timber being driven into metal, of an aircraft fuselage travelling 5,000 yards, a 20-ton railway coach going half a mile, and 'all 500 jobs lost' in one community with the total destruction of its mall.

Many Oklahomans stated that the subsequent disruption was not dissimilar to that resulting from the Oklahoma City bombing of April 1995. Policemen disagreed saying that 'the bomb was confined to one block, but the tornado's destruction just went on and on'. That 1999 tornado invasion was also not dissimilar to the event of 1991 when an equally numerous and violent set arrived, but most of the earlier venom was formed over rural areas. Consequently there were no deaths and only two injuries. Tornadoes can also occur in Europe and Asia, but are never so violent as those in the US. Should one develop in the British Isles, as can happen very occasionally, it is unlikely to be more than force 3 on that Fujita scale of 0 to 5. The British Midlands experienced two during the evening of 5 July 1999, and locals expressed amazement at the sight, but the twisters injured no one. The greatest damage was from rain which partnered the storm, notably at Selly Oak and Grantham, causing several streets to be flooded.

The tornado columns, so frightening and conspicuous a feature of these storms, become visible when the internal pressure is sufficiently low to cause condensation, or because dusty earth, and much else besides, has been sucked upwards to join the maelstrom. When hoardings, cars, bodies, trees and even trucks are added to the columns, these climbing higher and higher towards the blackened sky above, it is certainly no occasion to stand and stare, as with typhoons sucking beaches dry, but to head for shelter very speedily.

Nevertheless, along with the scientific investigators wishing to learn more about these disastrous vortices, there are also motorised tornadophiles – 'twister-chasers', 'storm freaks' – whose greatest thrill, season after season, is to drive as near as possible to the twirling, flailing, cavorting, lethal, and undeniably dramatic performances, with nature flexing its impressive muscles to show what can be done. The storm chasers even offer rides, via 'Tornado Alley Safari', 'Silver Lining Tours' and the like, to witness this muscular flexibility under professional guidance. Best of all, allegedly, is core-punching which involves driving right into a storm 'to get a better look'. What Dorothy did in *The Wizard of Oz* is plainly sufficient enticement for others, even if a safe arrival in Oz, or indeed anywhere, cannot be guaranteed.

The scientific purpose behind much of this chasing owes a lot to Howard B. Bluestein, a professional chaser from the University of Oklahoma for over 20 years. Generally regarded as the pioneer who made such hazardous work academically acceptable, he devised TOTO, the Totally Totable Observatory, named also after Dorothy's dog. It was used in some of 'Howie' Bluestein's earlier studies, and was incorporated into the film *Twister* which did for tornadoes what *Jaws* and *Titanic* had done for offshore swimming and trans-Atlantic ocean travel.

The problem, which is still far from being solved or understood, is why some major storms make tornadoes and others of similar power do not. Back in the 1970s it was even thought that cars might be a cause, with motorists passing one another on the right-hand side of each road initiating anticlockwise swirls of air. Now it is believed that the differences in temperature gradients may cause air to 'barrel' over the ground. If such a barrelling is twisted upright by the storm's updraught a tornado will then be on its way. That 1999 disaster was not good for people or property, but was excellent for scientists. They were able to collect more valuable information than ever before. As for the reason that some storms spawn tornadoes, and some do not, 'I think we'll have the answer in the next 10 years,' says Bluestein, chief scientific chaser, and author of *Tornado Alley*, an exciting and stimulating read.

Waterspouts are cousins but not brothers to tornadoes. The so-called tornadic spouts, formed in similar fashion to the land tornadoes, are rare. Far more frequent are those caused by a rotating quantity of air which is then associated with some form of updraught, probably linked with a developing cumulus cloud above, such as a congestus or a nimbus. In the northern hemisphere waterspouts revolve in anticlockwise fashion, as do hurricanes and tornadoes, but tend not to move north-eastwards, as do tornadoes, and are much less violent. They usually bend their path as they travel, their existence lasting for 15 minutes or so, and very occasionally for up to an hour. They are also impressive visually, but are nothing like so damaging as tornadoes. Waterspouts appear as if sucking water from the lake or sea over which they travel, but most of their visibility is due to water condensing in the lower pressures of their vortices. They can, and do on rare occasion, pick up fish, much to the astonishment – and gratitude – of land-dwellers when such very active and generous spouts are dying over land.

Waterspouts can occur singly, or in groups, and are commonest in coastal areas, although suspicion inevitably arises that sightings are more probable nearer land. Their diameter is about 100 feet, and their distances from water surface to cloud base are 2,000 feet or so. In theory waterspouts should only be possible when the sea is very warm, and most sightings do occur in such regions (although all night-time events go unreported), but they can also happen even in the colder months and colder times such as January. One particular and memorably big day for southern England was 18 August 1974 when several were observed cavorting in the English Channel.

Dust devils, or land devils, do not come into the same category as waterspouts, being dry rather than wet, and over land instead of water, but there are parallels. They are also upward-spiralling vortices of air that may reach 1,000 feet above the land and, typically, are about 60 feet in diameter. Best conditions for their creation are during the heat of the day with a calm prevailing wind, as strong winds cool the ground. They occur over arid areas, and it is dust which makes them visible, as well as unwelcoming when they cross one's path, rather than condensing water vapour. If they are particularly energetic they may even cause the formation of a cloud above them. This can give the impression that the whirling mass is emanating from the cloud, as with tornadoes, whereas the dust-storm cloud is formed as one result of the swirling updraught. Dust devils have no more resounding and scientific name, and do not truly merit a heading under Catastrophe, but have been known to flip roofs off houses and turn cars upon their backs. As for injuring humans they are content to fill eyes, ears and nostrils with quantities of sand, as well as fossilising upon the instant any exposed food about to be consumed.

Dust storms may be preceded by dust devils, but are more frequently initiated by a cold front moving across an arid area. They can certainly influence climate by diminishing sunshine at ground level, with dust storms earning the name if visibility is reduced to one kilometre/about half a mile, and are said to be severe if visibility is 50% or less of normal. The dust cloud of such storms can reach up to 10,000 feet and travel several thousand miles. New Zealand is often aware of storms from Australia, just as the US east coast used to be aware of dust from Oklahoma during the

1930s. Even Britons can wake up to find their cars speckled with grains from the Sahara.

Hail Less damaging in financial terms than hurricanes, but more destructive worldwide than tornadoes, this form of cascading ice can be impressively virulent. On 5 May 1995 one storm in the Dallas-Fort Worth area cost $2 billion in damage as well as injuring 90 individuals. The 15-minute storm which hit Calgary in 1991 caused $315 million worth of damage in that brief spell, a formidable $350,000 per second. Cars are particularly vulnerable; so too crops. It is not only the initial onslaught which can pulverise whatever is down below, but the fact that hail covers the ground temporarily with ice, thus killing much useful vegetation.

For some reason, or possibly for none at all, learned articles/paragraphs/dictionaries frequently forget formal measurements when discussing hail. They choose to use comparisons instead. In order of size, if I have got the order right, the dimensions of these rounded lumps of ice are compared with grapeshot, peas, grapes, golf balls, baseballs, tennis balls, pomegranates, softballs, oranges, grapefruits and fists, but I may be particularly wrong about grapeshot. When records need to be established it is necessary for rather more precision to come into play. For a long time Kansas held the prize with a 758-gram stone (melon-sized?), but this considerable giant was eclipsed in 1986 when stones of 1,020 grams hurtled to the ground in Bangladesh (with the width of poppadams?). Presumably it was necessary to make exceedingly speedy measurements of these freezing things, as every melted drip would lose another gram. Such hailstones would indeed hurtle, being compact and tipping the scales at 2.25 lbs. Bangladeshis have paid a grim price for being winners in this unpleasant competition. In 1986 a downpouring of such tremendous stones killed 92 individuals as well as great quantities of livestock. Allegedly a storm in 1888 killed 250 people 'in India', and that too may have been in modern Bangladesh.

Hailstorms, always formed in association with thunderstorms, are commonest in afternoons and evenings, as with tornadoes. Ice-producing storms tend to travel from south-west to north-east in the northern hemisphere. Therefore they are, once again, like tornadoes but unlike hurricanes. Most places in the world know hail, but only the warm temperate or subtropical land masses suffer the really damaging storms. In truly hot areas hail tends not to reach the ground, having already melted in the higher temperatures. Parts of central North America suffer the greatest incidence, as with tornadoes yet again. (The United States has, rather more than once, been called God's own country, but it receives a disproportionate quantity of devilish attention from catastrophic weather.)

Hailstones, whatever their size from pea to fist to melon and poppadam, are produced when super-cooled water is circulated within a thunder-cloud. This kind of water has a temperature below 0 degrees C., water's freezing point, and yet is still liquid. Contact with an aircraft can then cause 'icing' on its surfaces, a much feared happening, and contact with an already formed hailstone will lead to a bigger hailstone, with peas becoming grapes and so on. Large stones, if cut open, can show several layers, and 25 such rings have even been counted. These layerings of ice are formed when small stones encounter further regions of super-cooled water, perhaps by ascending and then descending within the cloud. For mere air to support solid objects the size of tennis balls, or bigger still, the forces and updraughts within a thundercloud become more compre-hensible. In time, with the hailstones expanding all the while, the cloud's updraughts will weaken, or the stones will grow too big, and gravity will then win. At once the grapes, baseballs, fists, and whatever plummet to the ground.

Hail showers occur in Britain, often in springtime when winds are either westerly or northerly, and more frequently in summertime. For any day to change, within very few minutes, from bright sunshine to falling ice and then to bright sunshine once again can be astonishing, whatever the size of the hail. A blessing for Britain is that its kind of hail is infinitely less punishing than in many warmer regions of the planet where this form of misfortune, along with hurricanes and tornadoes, is yet another species of calamity on weather's dark agenda.

Avalanches are quite unrelated to hail, but deserve a mention under the general heading of ice-cold catastrophe. The latter part of the 1998–9 winter in the European Alps was particularly devastating, not only for snowfall, but for the resultant avalanches which killed 64 in Austria, France, Switzerland and Italy. (Avalanche is from the French for a lowering.) Most severely hit was the Austrian community of Galtur where 31 died on 24 February 1999. Much later it was estimated that 100,000 tons of snow had descended in that avalanche, travelling at a peak speed of 140 miles an hour, before burying much of Galtur beneath 26 feet of snow. Small wonder, therefore, that buildings collapsed so readily when assaulted by such force, it being more surprising that anyone survived, as did a fortunate few. Plainly the normal snow-breaks, built here and there on Alpine meadows, can do little – if anything – to halt or slow down such a terrifying torrent of snow. Even faster speeds have been recorded.

Part of the problem has been the increasing popularity of winter sports, with 13 million skiers spending at least one night in Alpine resorts during 1980, and 60 million doing so only 15 years later. In fact the number of deaths-per-skier has fallen and the destruction of property has also lessened. Avalanche science has much improved, such as the careful observation and assessment of experimentally induced avalanches. This work has led to better zoning laws about new construction, and better warning systems for off-piste skiers. Even so there will always be more catastrophic years. The fact of an anticyclone becoming halted, by a complex assortment of winds, on the northern side of the European Alps during that winter of 1999 meant that snow continued to fall in the same region, rather than being spread more widely. This resulted in some areas receiving six times their normal quantity. Alpine records go back a long way, and 1999 has now been added to previous disaster years, notably 1720, 1808 and 1951.

All such extreme snowfalls can be hugely damaging should they lose their grip upon the mountainside. Avalanches destroy even with their advancing pressure waves before their snow arrives, but they do little to affect the planet as a whole and have not been increasing in recent years. It can merely seem as if they are doing so, particularly when they assault communities like Galtur which had been considered safe. That halted anticyclone was to blame, the kind of circumstance which has such vicious results every century or so.

Lightning Thunderstorms are happening so steadfastly they are not truly equivalent to the abrupt extremes of hurricane, tornado or pulverising hail. At any single moment a couple of thousand thunderstorms are occurring somewhere on the planet, with most of them in the damp tropics. Lightning is undeniably impressive, partly, or mainly, for the noise if it is close, and partly for the visual display. Huge energies are involved, many millions of horsepower, but these are usually dissipated harmlessly. People can be killed should lightning strike nearby, and four-legged animals are killed more frequently, not merely because there are more of them standing in the open but because their four feet cover a greater quantity of ground.

The tale is often told that giraffes are the most vulnerable animals of all to lightning strikes, but it is their four limbs and a tendency to stand in groups which increases their danger rather than their height. A phenomenon known as step potential is to blame. Every lightning strike raises that piece of ground's electrical potential. Therefore an electrical gradient is formed between that place and elsewhere. An animal or human standing near that point will have the nearer, or nearest, foot at a higher potential than the other foot or feet. Electric current will therefore flow from one foot to the other(s) and this current can be sufficient to stop the heart. Animals often collect in groups, as do humans, and this behaviour further steps up the step potential, increasing the current flow yet more damagingly from leg to leg to leg. For this kind of reason seven elephants in the Kruger National Park were killed in 1999 by a single strike. The best advice for humans, and for elephants, is not to stand in a group, but to crouch down independently and separately, with feet together, thus minimising the production of step potential.

Just as it is not precisely known why some low pressure systems develop into hurricanes, or why tornadoes develop as they do when apparently optimum conditions can also fail to induce them, the mechanism for creating lightning is one further feature not yet fully understood. Without doubt static electricity builds up within a cloud, being positive – it is thought – on ice crystals at the cloud's top and negative on the water droplets near the cloud's base. Simultaneously a positive charge will develop nearer to the ground underneath the cloud. For a time the intervening air acts as a successful insulator between the various charges, but a time is reached when this is no longer sufficient to prevent a discharge. Most of these take place between clouds, or between one part of a cloud and another, or between a cloud and surrounding air. Only a quarter or so of such discharges make contact with the ground, but this still means that the Earth is being struck, somewhere, 100 times a second.

Lightning can and does strike twice in the same place. The Empire State Building in New York City is struck about 500 times a year, and at the rate of once a minute if a storm is raging suitably. It is also possible, although rare, for humans to be struck twice. An American park ranger was struck eight times before shooting himself 'due to the pain' – and no doubt to the fear that he might be struck again. There is even a theory that a human body becomes more likely to receive a lightning bolt after each and every non-lethal strike. The body's physical properties are apparently altered, causing this extremely unfair loading of the dice.

Lightning is described as forked, sheet or streaked, but the terms are subjective, depending upon the observer's location and whether the strikes are obscured by cloud. North America, so prone to natural mishap, is particularly affected, with about 400 people struck by lightning every year and 100 dying as a result. The greatest lightning disasters occur when something critical is struck, such as explosives, or when fires are initiated. People also set fire to forests, but it is widely believed that about half of California's forest fires are caused by lightning. Some plants, notably in Australia, can only germinate when their seeds are involved in fire. As humankind has lived on that continent only for a short while in evolutionary terms, the need for fire must have arisen in response to natural lightning fires.

The temperature of a lightning bolt is in excess of 20,000 degrees C. The nearby air is therefore heated tremendously, causing expansion, and it then contracts after the lightning bolt has passed one 20,000th or so of a second later. Hence the crack of thunder, from the abrupt contraction, which may then reverberate and rumble as this noise bounces off the nearby clouds. In general the sound of thunder is no longer audible if created at a distance further than 20 miles, but its lightning cause can be visible for 50 miles or more, notably at night and over the sea. As sound travels at some 700 miles an hour, and lightning is virtually instantaneous, the storm's distance from an observer is approximately one mile for every five seconds between the lightning flash and its thunderclap. Lightning looks reddish if there is rain in the cloud and blueish if there is hail.

As almost every substance is a better conductor of electricity than is air the flash of lightning will make use of anything which reduces the insulating buffer of air between the negative and positive charges. Therefore it can involve any object which lessens the gap between ground and cloud, such as hills, trees, spires, buildings or even people and animals if they are convenient. Human hair standing on end suggests that a strike is imminent, in which case the person involved should crouch low and not lie full-length or spread-eagled. Standing beneath isolated trees, however protective from rain, is not a good idea, and two Thai women paid the price for such behaviour within London's Hyde Park in 1999. (The coroner investigating these deaths reported that metal in their 'under-wired bras' had acted as conductor, the second time in his experience of 50,000 deaths when he had encountered this phenomenon.) With lightning storms it is better to be inside a car, whose tyres provide insulation, or within a building where, if struck, the lightning will favour travel along either its plumbing or the wiring.

Does lightning travel upwards from the ground or downwards from a cloud? Both, because the negative 'leader' descending from the cloud attracts the ground-based positive charges, which leap upwards from the higher points. Benjamin Franklin was first to prove, in most courageous/foolish fashion, that lightning is a form of electrical discharge. The kite he flew during a Philadelphia storm did indeed attract electricity very forcibly, charging a key he had placed along the wire. The only mystery linked to this exciting experiment is a lingering curiosity why the exciting career of this inventor-diplomat did not cease at once. Certainly the next two men to try the same trick died when lightning struck their kites – and them.

Volcanoes Perhaps the wise and clever Franklin was even wiser and cleverer after that frightening brush with electricity. He may also have been a touch less brave subsequently as the lithographs portraying the famous kite occasion show him to be remarkably composed, and even unconcerned, at the very moment when a flash of lightning is lighting up his key. He did not make the experiment a habit, no doubt discouraged by those two deaths, but certainly continued his quest for atmospheric knowledge. For example, he subsequently observed how sunlight passing through a magnifying glass failed to ignite paper after an Icelandic volcano had erupted. Dust particles thrown into the atmosphere from that natural event had significantly diminished the intensity of solar radiation, and the scientific diplomat proved this point by his little, and non-lethal, experiment with a magnifying glass.

That Icelandic eruption was an extremely significant event, being (according to a 1995 paper by John Grattan and Mark Brayshay in the *Geographical Journal*) the 'most violent, extensive and prolonged volcanic episode' to occur in the northern hemisphere during modern times. Iceland's 'Laki fissure' started its activity on 8 June 1783 and did not shut down until 8 February 1784. There were five major surges, and the effects were widespread, notably in Europe, but they were most devastating on Iceland itself, where 79% of the sheep, 76% of the horses and 50% of the cattle died as a direct result of pasture destruction. An 'estimated 24% of the human population' also subsequently perished. In England the naturalist Gilbert White, so diligent in keeping records, referred to 'the peculiar haze or smoky fog that prevailed for many weeks'. The poet William Cowper wrote (more poetically): 'We never see the sun but shorn of his beams, the trees are scarce discernible at a mile's distance, he sets with the face of a red hot salamander and rises with the same complexion.'

Over in Paris, serving as US Ambassador, Benjamin Franklin wrote of the fog being 'of a permanent nature; it was dry, and the rays of the sun seemed to have little effect towards dissipating it, as they easily do a moist fog arising from water'. Newspapers of the day did not customarily detail weather, but broke with tradition in that summer of 1783. 'The weather has been very remarkable here; a kind of hot fog obscures the atmosphere' – *Aberdeen Journal*. 'The fog is not peculiar to Paris, those who come lately from Rome say it is as thick and hot in Italy, and even the top of the Alps is covered with it . . . travellers from Spain affirm the same of that kingdom' – *General Evening Post*. 'In some places the fog withers the trees, and almost all the trees on the borders of the Ems (in Germany) have been stripped of theirs' – *Ipswich Journal*. It was a great occasion for happy phrases – 'uncommon gloom', 'crispens the turf', 'sulphureous stench', with the sun 'like a pewter plate red hot'. Some priests, quick to discover meaning in the fog, made reference to the Day of Judgement. There were also tremors that year, causing a Devonian vicar to write in the *Exeter Flying Post*: 'When the Lord ariseth to shake the earth, it may be presumed, I hope, that the inhabitants thereof will learn righteousness.'

What they were not learning was the reason for the prolonged, 'violent', 'tremendous', 'dreadful', 'remarkable' and 'portentous fog'. Not until 1784 did Franklin suggest that the 'strange phenomena' might have been caused by the eruption of an Icelandic volcano. By then the Laki fissure had concluded its business, and the sun was shining brightly once again. That Icelandic event was undoubtedly memorable, but not unique. Volcanoes are often all too liberal in their effusions, with ejected lava and rocks intermingling with ash and dust. Sometimes their effluent is almost entirely dust and, on other occasions from other volcanoes, little more than lava pouring from a vent. The heaviest material spewed upwards will fall at once, perhaps to join the lava, and much of the dust which only reaches the lower atmosphere will soon be washed to earth by rain. It is the higher dust which causes the sun to be shorn of its beams, if not into a red hot salamander.

The most generous eruption of the 20th century, in that it killed an entire town in Martinique, occurred on 8 May 1902. That date was a public festival and many had crowded into the urban area to enjoy the celebrations. Mount Pelée, largest mountain on the island, which loomed (and still looms) malevolently over Saint-Pierre, much as Vesuvius does over Naples, poured forth a pyroclastic flow as from 7.50 am. This hurtled down the slopes, far faster than anyone could run, and wiped out that island's capital city. All 28,600 of its people died, save for the murderer Augustus Cyparis and one other who had been safely and deeply ensconced in prison at the time. (Cyparis was badly burned but received a free pardon before joining a travelling circus and earning a living describing his miraculous escape.)

What Franklin was observing in 1783, or rather not observing save for their effect upon sunlight, were the masses of microparticles, the several cubic miles of matter from a major eruption which can reach the stratosphere. These individual specks are so small, being a micrometre or less in size, that their falling speed is pathetically slow. They can therefore remain aloft for years, circling the Earth and darkening the sun, thus preventing scientific diplomats from setting fire to paper. Such dust veils, as they are known, have recently resulted from El Chichon's eruption of 1982, from St Helens in 1980, when 1,300 feet of the mountain's top was blown skyward causing ash from that explosion to reach an altitude of 12 miles, and from Pinatubo in the Philippines in June 1991 which injected some 20 million tons of sulphur compounds into the atmosphere plus great quantities of dust. It is thought the low global temperature of 1992, as against the warmer years of 1990, 1991, 1994 and 1995, was due to all that dust, a suggestion that Franklin would have been happy to promote.

The very greatest eruption in recent times, taking place at Tambora on the Indonesian island of Sumbawa in 1815, caused that year – even in distant Europe – to be known as the 'year without a summer'. Napoleon would agree with that, it being the year of Waterloo, but even the following year, when he was safely housed on the island of St Helena, was also cold and dismal. The biggest eruption in the past 100,000 years occurred at Toba in present-day Sumatra. It is thought to have been 'at least' five to ten times the size of Tambora, and happened to coincide with the cooling climate which preceded the last ice age. The dust surely hastened that cooling, causing not so much a year without a summer as a period even colder than it would otherwise have been. There is no reason, of course, why another such devastation should not occur right now.

No one at the time of Tambora knew what to blame for the poor weather, and no one in Europe even knew for quite a time that anything exceptional had happened in Indonesia. The mountain had erupted on 10 April in that soon-to-be summerless year of 1815, removing almost 5,000 feet from the volcano's height above sea level. Only during the following November did *The Times* of London publish a report, this having been sent by an East Indian merchant: 'We have had one of the most tremendous eruptions that ever perhaps took place anywhere in the world.' Much like the apocryphal 'Small earthquake in Chile, not many dead', no one in London, or indeed anywhere save for Indonesia, cared much that a merchant, in reporting the incident, had provided this item of news before adding that his ship's progress had been much impeded 'for many miles . . . (by) trunks of trees, pumice and etc.'.

No one began to put two and two together, about bad summers and distant volcanoes, until Krakatau blew its top in 1883. And no one indicted Tambora for the climate of 1815–16 until ice cores recently collected from Greenland showed a layer of acidic ash which had been deposited precisely at that time. A Danish glaciologist working in the 1970s was first to point the finger of blame at the catastrophe which had hit the island of Sumbawa a century and a half earlier. He was able to link the icy ash with the far-off place it had come from. It is now believed that Tambora's explosive force was equivalent to a billion tons of TNT, as against Hiroshima's 20,000 at the end of World War Two.

Indonesia is particularly susceptible to destruction by volcano, possessing the greatest active concentration. Almost half of its volcanoes have erupted in historical times. The uninhabited island of Krakatau, midway between the Indonesian islands of Java and Sumatra, experienced not so much an eruption as a detonation. On 27 August 1883 four cubic miles of rock, or maybe four times as much, say some vulcanologists, were blasted up to 30 miles within the stratosphere, and barographs all over the world recorded pressure changes. People also noted the above-average excellence of sunsets for two years after that happening, these formed by the prevailing dust. It is even postulated that some of Joseph Turner's paintings reflect that colourful time. The Krakatau event has since been used as a baseline for what is known as the Dust Veil Index, with the Krakatau DVI set at 1,000. The 1815 eruption from Mount Tambora is reckoned to have scored 3,000 on this scale.

Such massive injections of dust not only obscure the sun but also lower the temperature, with the year after Krakatau being the coolest on record between 1880 and the mid-1990s. In general each large eruption lowers the all-round global temperature by about 0.5 degrees C., but not necessarily. El Chichon's eruption, the Mexican event of 1982 which occurred when countless thermometers were in place worldwide, did not cause any drop, although such a fall had been widely anticipated after the eruption. Perhaps Chichon's horrendously lethal fury had helped to stoke that particular expectation, it being the deadliest eruption, with 1,800 killed, ever known on the North American continent. If humans are indeed adjusting terrestrial temperatures, as is generally believed, they are being influential on a par with single major catastrophes such as Krakatau.

As for mortality resulting from eruptions, a major cause of death-by-volcano occurs because one generation's disaster can be the next generation's livelihood. When lava and ash have been sufficiently eroded by weathering they form exceptionally good soil. This substance usually falls short of being the complete natural fertiliser only by a deficiency in nitrogen, a shortage easy to remedy artificially. Therefore people do like living in volcanic areas, within the shadow of the mountain which is both blessing, for the soil it has created, and disaster for the horror which may ensue.

The unpredictability of eruptions makes this gamble attractive and worthwhile, notably for those who cannot afford land elsewhere. Living with volcanoes is also attractive for vulcanologists and others who wish, albeit riskily, for a hands-on approach to their work. When Mount St Helens, Oregon, was threatening activity in 1980 the scientist Harry Glicken was stationed six miles from its peak at the Coldwater II observation post. On 18 May he was replaced by David Johnston who, a few hours later, was killed by the eruption.

Glicken himself was then killed 11 years later by a pyroclastic flow on Mount Unzen in Japan, together with an assortment of film-makers and journalists. Such flows, hot and liquid, had so often coasted down one particular Unzen valley that 43 individuals, including Glicken and other scientists, were standing on the valley's upper wall on 3 June 1991 to watch the famous sight. Unfortunately the flow that year proved to be so tremendous, and so fast, that it engulfed the entire valley, killing all those bystanders instantly. Such flows of pulverised magma can indeed be speedy. Two men observing the St Helens eruption of 1980 saw, only too clearly, a flow suddenly coming their way. They leapt for their car and drove at 90 miles an hour for 12 miles before they could relax.

Nature is also swift, in general, to restore a situation which nature has destroyed. On Krakatau a solitary spider and a few blades of grass were found on the island one year after the explosion that had reduced the island to one-third of its previous size. These invasions were therefore occurring despite some 200 feet of ash covering what remained of the island's former bulk. Plant communities on the coastline were the earliest to become properly established, and ferns and grasses were first to exploit the interior, this region later becoming forest. A formal three-day search, initiated 25 years after the explosion, discovered 202 species of non-microscopic animals. A further expedition in 1919 found 621 such species during a search lasting 16 days. Yet another foray found 880 species in 1933, half a century after there had been nothing on the island, save for devastated rock and ash. Krakatau, also known as Rakata, has still not reached its former status in species numbers, possibly for being so much smaller than it used to be. It now has 300 plant species, a total less than formerly existed, and 30 species of bird, with 50 having been noted there at some time in all the years before the disaster of 1883.

Together with the approximately 600 active volcanoes on Earth, there are about 10,000 extinct volcanoes of any size, the numbers being necessarily vague because dead volcanoes can come to life, as Vesuvius did so awesomely in AD 79, before burying and immortalising Pompeii in ash, and volcanoes still suspected of being active can become inactive, although generally given the benefit of doubt and presumed to be biding their time. Vesuvius certainly has not ceased its activity, with important and recent eruptions having occurred in 1873, 1906, 1929, and the wartime year of 1944. Carefree bomber pilots, with nothing better to do, were known to have dropped high explosive into its crater which may, or may not, have hastened the 1944 eruption. It is even alleged, by some, that official policy encouraged such targeting, this intended to be a demonstration that the invading forces were all-powerful.

A few eruptions can arise from nothing whatsoever, as did Paricutin in Mexico in 1943. A farmer, busy ploughing, was amazed to see some smoke arising from a furrow. Soon modest quantities of rock and lava were being showered from that smoke's little hole. On the following morning there was a cone 25 feet high. One week later this was 500 feet, and after a further two months it had reached 970 feet. After two years, having reached a height of 1,390 feet, it then had a base over 1,000 yards in diameter. (No one, in retelling this tale, describes the ploughman's fate. Did he give up agriculture altogether, keenly aware of its awesome improbabilities?) Overall there are about half a million square miles of Earth's surface covered with fairly recent volcanic material, including the relatively modest quantity which now sits on a half-ploughed field in Mexico.

The title of a 'big one' is normally reserved for the next Californian earthquake, but it could also apply to the next really tremendous volcanic eruption. Mt Pinatubo's effusion was undoubtedly big, in that over two cubic miles of material were belched forth within five hours to bury the surrounding landscape beneath some 650 feet of ash, but something very much larger is always possible. At Yellowstone, a mere 600,000 years ago, 250 cubic miles of material were blasted forth, this event shrinking the Pinatubo effort into a pop-gun against a cannon. That single Yellowstone effusion engulfed large parts of three US states and despatched dust to cover most of the country. A group of fossilised mammals, including early rhinoceroses, was found recently in Nebraska 1,000 miles from Yellowstone. It appeared as if they had died from some pulmonary, and possibly ash-related, problem. There are no volcanoes in that state and it is believed (by some) that these animals died as a consequence of the ash from north-western Wyoming. The datings of the two incidents – dead rhinos and volcanic effusion – are similar.

Despite intensive searching no one could find the Yellowstone caldera, the crater formed by the collapse of the cone, until NASA happened to conduct an airborne test using infrared equipment in that area. A circular outline then became extremely visible, this shape 43 miles by 18 miles which therefore encompassed almost the whole of Yellowstone National Park, its outline being impossible to detect from the ground and also impossible from the air without such infrared equipment. Judging from local ash deposits there seem to have been three similarly cataclysmic eruptions from the same source, thought to have occurred 2 million, 1.2 million and 600,000 years ago. This regular incidence suggests that another such event may be due. Moreover a lake in that area has tilted at one end by 29 inches since proper surveying was first carried out (in 1923). It would therefore seem that the surface shape is changing, awesome indication that magma may be accumulating some eight miles deep, perhaps in preparation for a fourth and equally devastating effusion.

The Yellowstone event came from what are being known as super-volcanoes. The 15 or so ordinary eruptions each year, whether big or small, are in a different class. The super kind are greatly larger and greatly rarer, with the last of them occurring in northern Sumatra some 74,000 years ago. This site is now occupied by Lake Toba, and that caldera's dimensions are 62 by 37 miles. The very recent effusion from Mt Pinatubo lowered temperature globally by half a degree C. It is thought that dust from Toba was about ten times more effective, lowering the Earth's warmth by five degrees. Currently there is no way, for our peace of mind, that these detonations can be predicted. There is also no way of knowing how many of us might survive, or did survive last time. Planet Earth could still be the death of us, even though it has permitted us to evolve and live – thus far. Some calculations based on mitochondrial DNA and its traditional mutation rate seem to suggest that only a few thousand human beings survived the Yellowstone event.

Hurricanes, tornadoes, waterspouts, hailstorms and lightning bolts are all extreme manifestations of climate. Volcanoes, however extreme, can be different, quite frequently being the cause of changing climate rather than its effect. They are therefore more equivalent to human interference. The big eruptions (and not just the super variety) do affect the atmosphere, much as people are doing, and may even affect it more than people are yet achieving, particularly if another major upheaval comes our way again, something even bigger than Krakatau or Tambora. The comparison might make people think that man-made and unnatural interference is as nothing when set beside such natural might. In a sense that is true, with cataclysms so frequently and so damagingly making the point, but the thinking is wrong. Volcanoes do alter climate, and people may also be doing so in an equally unwelcome fashion. They can each disturb the normality on which we all rely.

It is also normal, however regrettably, that Earth's surface occasionally shudders, much as it has done since the beginning of its time. The tremors are a further form of catastrophe, not intimately linked to climate change but part and parcel of our planetary life. They cannot be denied.

Earthquakes These occur around the globe with a similar distribution to that of volcanoes, but neither causes the other although they can accompany each other. They both arise from the same general conditions: either there is a folding and a faulting of the crust or, to a lesser extent, there used to be such disruption. They usually occur (85%) in the top 40 miles of the Earth's surface, but may be between 40 and 200 miles below it (12%) or even deeper (3%). Although an earthquake's point of origin or focus can be deep or shallow its epicentre is always said by seismologists to be the place on Earth's surface lying directly above that point of origin.

Essentially earthquakes are a sudden release of energy, this accumulating when rocks can no longer slide easily over each other. Thousands of earthquakes occur each year, with most being minor affairs detectable only by seismic instruments. The distribution pattern of active epicentres is complex but straightforward. In the New World it ranges from Alaska down the entire west coast of both Americas, with a bulge around central America, including the West Indies, and a south-westward extension from that middle belt out into the Pacific. There is another major belt running centrally the entire length of the Atlantic Ocean. This then bends a long way round southern Africa before heading north-east of Madagascar and then more westerly to meet Africa at its horn, thus linking with the famous Great Rift which has caused so much of East Africa's spectacular scenery. A further belt travels eastwards from the mid-Atlantic through the Mediterranean, along the North Anatolian fault in Turkey, where 15,000 died on 17 August 1999 from the 7.4 Izmir quake, and swells considerably east of the Caspian to the Himalayan region and above. The final, and most infamous, belt travels down eastern Asia, involving Kamchatka, Japan, Borneo, Indonesia, New Guinea, parts of the southern Pacific, and then on through New Zealand. Tremors can also be felt, and damage can occur, outside of this general distribution. Even Britain has experienced a few earthquakes, although these are incomparable with major events elsewhere. Inverness, near to the Great Glen fault (which slices through Scotland and continues up past Caithness and Orkney), experienced an earthquake in 1901 and another, although less damagingly, in 1934.

Seismographs, or seismometers, make use of an arm pivoted at one end with a weight hanging from its free end. The instrument itself moves with the moving ground, but the lightly suspended weight remains, temporarily, in place owing to inertia. A mirror on the end of this pivoted beam reflects a ray of light on to a rotating drum, thus registering the weight's eventual movement (or movements as tremors do not come singly). Three seismographs are customarily used, one recording north-south movements, another east-west vibrations, and the third every vertical displacement. When all three forms of measurement are inspected, and allowances have been made for the speed of transmission through rock, the earthquake's epicentre can be assessed, along with its degree of intensity.

Together with the three instruments, each set in a different plane, there are three kinds of vibration for them to measure. These all travel outwards from each earthquake's point of origin, but at different speeds. Fastest, and therefore first to arrive elsewhere, are the compressional or longitudinal waves. Next are the transverse or shear vibrations which are slightly slower. Third are the surface waves, transverse vibrations with a large amplitude. These may be slowest but they do the greatest damage.

Two forms of classification are used to measure earthquakes. The Modified Mercalli Scale (MMS), graded from 1 to 12, measures property destruction, ranging from 1 (negligible tremors, detectable only by instruments) to 4 (felt indoors by many) to 7 (everyone runs outside, weak buildings are damaged) to 10 (some landslides, many buildings destroyed) and finally to 12 (catastrophic, considerable damage everywhere, and ground warped).

The more famous Richter scale is less immediately comprehensible, being logarithmic and also open-ended. C.F. Richter, who died in 1985, devised it in 1952, promptly causing his name to become a household word. His scale's principal virtue results from being based upon objective recordings made by the standard forms of seismometer. Magnitude 0 is 8 billion ergs, with one erg (a unit now rarely used) being a minute quantity of energy or work done, whereas magnitude 9 on Richter's scale is 1.2 ergs multiplied by 10 to the power of 25, or 12 million billion billion ergs. Richter 2 is about equivalent to MMS 5, and Richter 7 is roughly equal to MMS 10.

A major Richter advantage is that seismologists, solely by looking at their instruments, can know the severity and location of an earthquake without leaving their laboratories. They can therefore guess/assess the scale of destruction. MMS numbers, on the other hand, can only be attributed after the location has been visited and its damage has been inspected. The smaller Richter quantities, such as 3 or 4, define earthquakes of little consequence. Anything over 7 is likely to be a disaster, and certain to be one if there are people and houses in that area. The biggest quakes, scoring 8.0 or more, are almost always horrendously destructive, with 8.6 being the highest so far recorded.

For any given earthquake there may be little comparison between its MMS and Richter definitions because property destruction does not necessarily correlate with an earthquake's actual magnitude in terms of its energy release. Structures built on sand or clay suffer more severely than those on rock. Similarly modern buildings constructed with earthquakes in mind are less likely to fall down than old-style single-storey dwellings. Iranian and Afghan earthquakes, for example, are often particularly lethal, with so many of their traditional homes made from bricks of dried mud. These fall, alas, with the greatest of ease, burying their occupants beneath the weight of the construction material.

Japan, only too aware of earthquakes, having experienced 17 in the 20th century with magnitudes greater than 7.0, is understandably desperate to prevent the scale of earlier catastrophe. Worst of all was the Great Kanto quake of 1923 which, with a magnitude of 7.9, killed 142,807 people and damaged or destroyed 576,000 houses. Such a single act of destruction is hard to visualise, although the Kobe quake of 1995 did its best by reminding everyone what an earthquake can do even in modern times. It had a magnitude of 7.2, killed 5,504 people and damaged over 100,000 houses. It certainly caused architects and engineers to look again, both at their theories and their calculations.

In one sense this most recent major catastrophe was more appalling than the Great Kanto devastation of 80 years ago, in that innumerable measures had been taken since that time for safer houses, better bridge construction and firmer building of all kind. The Japanese people had also become better educated about an earthquake's possibilities, with so many tremors, from unimportant to devastating, constantly serving as reminders. Within the last century, and totalling the effects only of the major quakes above Richter 7.0, this one nation has lost 165,000 of its citizens and 720,000 of its buildings, or an average of 5 people and 20 of its buildings every single day. (Nevertheless traffic in Japan is currently killing about four times as many people every day, again on average. It used to kill over eight times as many until Draconian laws were introduced to cut down the toll.)

China, with so many more inhabitants, frequently swamps statistics with the scale of its catastrophes. A 7.6 earthquake which hit Tangshan in 1976 killed 'more than 200,000'. Ancient building techniques were blamed for this tremendous number. In the United States, despite so much construction along the famous San Andreas fault, which runs through much of California and is clearly visible from the air, there has not yet been a rival, in terms of death and destruction, to the San Francisco disaster of 1906 which killed 3,000 and laid waste great sections of that city. A major quake in southern California killed 700 in October 1911, this being a great disaster with a major loss of life, but the deaths were still less than a quarter of San Francisco's total.

That event of 1906 ruptured gas mains which spread fires throughout the damaged area and added to the horror. The motion of the ground on that day was mainly horizontal, with one side of the fault moving a maximum distance of 15 feet relative to the other. No wonder that gas mains broke apart, buildings fell down, and so much death occurred. The next 'big one' is anticipated, some form of repetition of that earlier event, particularly at the southern end of the 700 mile-long fault where major stresses are now accumulating. This looming Californian earthquake is not necessarily imminent, but it is inevitable and it will devastate.

Around the world, and within the single decade of 1980–90, some of the major earthquake tolls have been:

Armenia 1988 (100,000 dead);
China 1988 (1,000);
India–Nepal border 1988 (500);
San Salvador 1986 ('scores');
Mexico City 1985 (10,000);
Turkey 1983 (2,000);
Southern Italy 1980 (3,000);
Algeria 1980 (20,000).

These eight major tremors therefore killed at least 137,000 people in a mere ten years.

Casualty numbers from earthquakes are formidable, and presumably always will be, notably in the poorest areas, but the history of humankind would not have been greatly different, or will not be, even if the numbers were or will be ten or more times greater. As with the other catastrophes, when nature lays waste particular areas, they only cause a hiccough in development, much like rocks thrown into streams which temporarily disrupt the flow. Krakatau, one century after its near-absolute destruction, is now alive and well, rich with life of every kind, although not quite so rich as previously. Earthquakes are not as direct in their effects upon the climate as are the atmospheric perturbations, like hurricanes and thunder-storms, but they can make alterations. They disrupt water flow, they often affect the landscape, and the previous patterns of vegetation can in consequence be changed quite dramatically. As a result there are climatic knock-on effects, particularly locally.

There is no universal signal of an impending earthquake, but an increasing intensity of minor shocks serves, generally, as presage to some major seismic event. Japan's seismographs normally record about 7,500 shocks a year, or some 20 a day when nothing much is happening. On 20 November 1930, in a famous and well-recorded incident, there was a sudden surge to 100 shocks. This number continued to increase and, by the 24th, had risen to 700 such disturbances. Finally, on the 25th, there was an earthquake, but no one had been able to foretell where its damage might occur, or how great the shock might be, or how much hurt it might inflict. In fact 300 were to die in Mishioma.

Towards the end of October 1998 a geologist from Edinburgh University warned Icelanders to expect either a Richter scale 5 earthquake in the 'near future' or a more severe disturbance before the end of February 1999. Stuart Crampin issued his final warning of a 'quake's imminence' on 10 November, this prophecy confirmed three days later by a Richter 5 event. It did no serious damage, and there was more subsequent interest in the anticipation. Might earthquakes become predictable? If so, posing a greater problem, might their precise location be forecast with any degree of precision? Southern California is expecting the 'big one', with 'when' still being one unknown and exactly 'where' the other.

An even bigger catastrophe, with even greater need for early warning, may well occur in several southern republics of the former Soviet Union. In 1911 an earthquake blocked the River Murgab in the Pamir mountains. It is now believed that another earthquake in that region might act contrarily to cause an unblocking of that river. Such an occasion could kill hundreds of thousands of people. The Sarez lake in Tajikistan, created by the earthquake and by its blockage of nine decades ago, has been filling ever since. It is now about 40 miles long, 1,600 feet deep, and contains over four cubic miles of water. If an earthquake were to release that water, perhaps by a landslide setting the process going, it is believed that a wall of water over 300 feet high would promptly rush down the narrow Murgab gorge. Even after 600 miles it could still be 20 feet high, and some 19,000 square miles of land might be flooded. That land, in Tajikistan, Uzbekistan and Turkmenistan, as well as the northern part of Afghanistan, is home to five million people. Hence the possibly horrendous loss of life should nature choose to send another quake to undo what it did 90 years ago. Therefore a desperate need exists for some kind of downstream warning system, should it ever be anticipated or actually be seen that a monstrous wave of water is on its way. Ironically this water would eventually reach the Aral Sea, the region east of the Caspian so desperate for refreshment, having lost half its volume in the past 30 years. The place would surely prefer a gentle inward flow rather than a single, and no doubt devastating, deluge.

Landslides can occur even without earthquakes to set them going. Excessive rain can cause some slope of land to lose its grip upon the rocks beneath, and then for tons of earth to slither down the hillside. The kind of damage which ensues can be either minimal or horrendous. In March 1963 such a landslide careered downhill to arrive at a man-made lake in Italy, this reservoir kept in place by a concrete wall of tremendous height within a narrow gorge. The impact from the slide was so tremendous that thousands of tons of water were sent cascading over that concrete wall. This torrent then hurtled down the valley, killing and drowning 1,909 individuals as it did so. There was no earthquake, and no failure of the concrete, but there might as well have been for the slaughter which occurred.

Mere rain can often be devastating when it falls exceptionally heavily and is then channelled into a narrow valley. Lynmouth in England suffered tremendously one day in 1952 when eight inches of rain fell on the moorland up above. Much of it then cascaded down two steep valleys to assault the tourist resort, killing 34 as it did so. An Italian community built within sight of Vesuvius was untroubled by that volcano until May 1998 when the town was horrendously damaged by a mudslide arising from that mountain's slopes which poured downhill to hurtle through streets at 50 miles an hour. Damaged individuals were taken initially to the Sarno hospital, but that too then succumbed when additional mud arrived. A total of 160 died, 15 of them in that hospital. Similarly six inches of rain falling in 45 minutes above a Spanish camp site in 1996 also led to devastation. Not a drop had fallen on the site itself, but 86 were to die when the sudden rush of water from above made short work of cars, caravans and people in that exotic setting.

'There has never been anything like it before,' say those who have lived in such disaster areas all their lives, but they can be wrong. At Lynmouth, after diligent archive research following the disaster, it was learned that such flooding had occurred every 150 years or so. What is new is the belief that modern precautions, with concrete sluice-ways and the like, will be sufficient. What is also new is the pressure to build in vulnerable regions, in places where – if heavy rain and a mudslide do occur – the effects will be lethal. The final novelty is that we all get to hear about disaster, whether the happening is in North Devon, central Italy or Uzbekistan. We also tend to conclude that climate change, whether man-made, man-encouraged or entirely natural, is becoming worse.

Tsunamis Better warning is feasible with the wrongly called tidal waves – more correctly known as seismically generated sea waves – which cross oceans at phenomenal speeds, on occasion to cause havoc in other continents than those of their creation. Tsunamis are formed by earthquake activity below sea level, or by underwater landslides or volcanoes, and their subsequent speed is proportional to the depth of the water through which they pass, being faster when it is deeper. Ships' sailors do not, in general, even know or notice that a tsunami has passed beneath them. (The word is Japanese, meaning harbour-wave, and has been universally adopted.) Their wave height in deep water is three feet or even less. Ordinary ocean waves travel in leisurely fashion, but these tsunami waves, which have nothing to do with tides, can speed along at several hundred miles an hour. They are therefore pacing, more or less, the airliners flying high above them. No wonder that sailors do not notice any tsunamis which chance to come their way.

When reaching shallower water they slow down to some 30 miles an hour. They then rise greatly in height, to 50 feet or so, and finally vent their considerable energy upon the coastline which happens to be facing them. The massive energies involved have been forcibly squeezed by the shallower water into a smaller volume, and the tsunami height grows tall in consequence. (Ordinary waves, when breaking on the seashore, behave in similar fashion, rising steeply as water depth decreases.) Krakatau's volcanic explosion in 1883 produced fearsome seismic waves, allegedly over 100 feet high on reaching land. This island lies in the Sunda Straits directly between Java and Sumatra. Waves from its awesome detonation killed an estimated 36,000 individuals on those two other Indonesian islands. These people first heard the explosion, which was audible for 600 miles, and then died several minutes later when the waves assaulted them.

Warning of this form of catastrophe is now possible because seismographs and tide gauges can provide the necessary information. A recent development involves the combination of surface buoys, of sensors on the sea floor, and the global positioning system of navigation satellites. When a sensor detects a tsunami passing in its locality it transmits information to the nearby buoy which then alerts a satellite. Tsunami waves are detectable by their very long wavelengths, these – in deep water – being several hundred times greater than ordinary ocean waves. Vulnerable regions are then notified and, despite tsunami haste, there is some time for evacuations to be effected. Unfortunately most tsunamis are destructive fairly near the coastlines which were their point of origin, giving little time for warning.

If an event off South America triggers such a wave there are several thousand miles of Pacific Ocean to be crossed before this concentration of energy flings itself upon a Japanese or Hawaiian beach, and at whatever seaside buildings lie within reach of this form of natural rage. One tsunami, plus an accompanying storm, killed 1,000 in the Bay of Naples in 1910, and another drowned 3,000 on the Japanese island of Kiu-Siu in 1927. Hawaii suffered 159 deaths in 1946 when a tsunami arrived from an Aleutian earthquake of Richter 7.8, whose epicentre lay 2,300 miles away. The Hawaiian community of Hilo, with a seafront facing north and therefore susceptible to tsunamis heading south, suffered 96 of those 159 casualties. Five Coast Guard employees had already died when that same tsunami had entirely destroyed the Scotch Cap lighthouse in Alaska, and total damage from this single destructive wave was $26 million.

The Pacific is particularly prone to tsunamis, with 86% of them being the result of undersea earthquakes around its rim. An earthquake 100 miles off the Chilean coast in 1960 caused a wave of sufficient power to kill 119 when it reached Japan. Tsunamis have not been as lethal for that vulnerable nation as earthquakes, but are rightly feared. Papua New Guinea is not normally so beset by these violent events, but will not forget 17 July 1998. Shortly after sunset a nearby earthquake of Richter 7.1 sent a wave 25 feet high crashing on to Papua's northern shores. Some 2,200 villagers then died, with 30 more of them subsequently losing injured limbs to gangrene. One survivor, having heard a rumble 'like a low-flying jet', was then washed 1,000 yards inland. He had tried to run, but the wave had overtaken him.

The 1990s decade was severely hit by tsunamis, with ten of the most destructive on the Pacific Ocean's boundaries leading to almost 4,000 deaths in that time. *Scientific American* has listed these 'killer waves', together with their maximum wave height and the number of fatalities.

2 September 1992	Nicaragua	32 ft.	170 dead
12 December 1992	Flores I., Indonesia	85 ft.	1,000
12 July 1993	Okushiri, Japan	100 ft.	239
14 November 1994	Mindoro I., Philippines	22 ft.	49
9 October 1995	Jalisco, Mexico	36 ft.	1
1 January 1996	Sulawesi, Indonesia	11 ft.	9
17 February 1996	Irian Jaya, Indonesia	25 ft.	161
21 February 1996	North Peru	16 ft.	12
17 July 1998	Papua New Guinea	50 ft.	2,200

In that ten-year period of the 1990s there were 82 reported tsunamis in the world, a higher figure than the average total of 57 per decade. The number of waves is thought to be increasing, partly through better communications. The death toll is probably increasing because more people are living in the vulnerable coastal areas. Although the long-distance tsunamis are impressive, particularly when they cross the entire Pacific Ocean, it has been estimated that 90% of all tsunami deaths are due to undersea earthquakes within 125 miles of the spots where these waves have speedily created their particular form of natural catastrophe.

Fire Also known as bush fires (Australia), scrub fires and forest fires, wildfires – the proper generic term – not only kill, but destroy, regenerate, are often costly, and certainly affect the climate. Among the most lethal, the community of Matheson in Ontario experienced 244 deaths in 1916. The most expensive, to date, was the fire which hit Oakland, California, in 1991. It consumed 3,300 homes and cost $1.5 billion. Mexico, the United States and Canada experience 120,000 wildfires each year, with the majority in the US.

Climate provides the right conditions for fires – drought, hot temperatures, dry winds – and climate can also set them going, with thunderclouds frequently providing the initial spark. Humans, it is thought, are less guilty, with arsonists not necessarily choosing the optimum time of heat and drought. Smoke from the fires can then affect the weather, its minute particles encouraging water to condense, with the subsequent clouds then reducing the quantity of sunshine arriving at the ground. Some recent work has shown that smoke from forest fires can even prevent clouds from forming raindrops. Its particles do cause water to condense, but the moisture is then divided among so many droplets that they are too small to fall as rain. It is even thought that fires are partly responsible for a decline in tropical rainfall.

Wildfires can certainly form clouds directly, lifting water vapour as they burn to form so-called pyrocumulus. If these puffy exuberances, often seen above Amazonia's biggest fires, grow big enough, and tall enough, they may eventually become the source of rain, thus actually helping to extinguish the conflagrations which gave them birth. Or, as extra anomaly, the pyrocumulus may grow to become a towering thundercloud, perhaps then producing lightning to create more flame.

As further contradictory point such fires are often beneficial as well as destructive. Ecologically they create ash, rich with nutrients for regrowth. Some seeds, as from various pines, germinate best, or only, when heat has opened up their cones. Trees tend to suffer from combustion more than grass, and certainly take much longer to recover their former status. Consistent grassland fires keep an area from becoming forest, with the potential giants all killed before they can dominate. This maintenance of grassland amends climate because the two forms of vegetative cover, dark forest versus open prairie and savanna, respond so differently to sunshine, to the retention of water in the soil, to evaporation, and to the presence or absence of rain.

Humankind has long been influential in this regard, burning to improve the grassland, burning for driving prey, or burning to create more open land for farming. Our ancestors were therefore affecting climate long before our modern times. Fires for warmth, for cooking, for keeping animals at bay were as much a part of human evolution as weapons and as speech. Either in error, when strong winds blew, or deliberately to clear land for a multitude of reasons, the hominid possession of fire gave even early humans the ability to amend the planet on which they lived. This power has undoubtedly accelerated in recent years, but does have an ancient history. Adjustment of climate may only have been modest in the early days, but it has steadfastly partnered mankind's development, notably through the use of fire.

To sum up. Weather is what people expect, the climate being broadly similar from year to year. Catastrophe, in whatever form, is exceptional. It may be comprehensible, in that hurricanes, for example, do form quite regularly in the warmer latitudes of the North Atlantic every summer and autumn, but the places they attack, the power of their winds, and the duration of their savagery are all most variable. They therefore arrive, along with tornadoes, hail, flood, earthquake, tsunami and wildfire, as sudden and extra hazards to the vagaries of weather.

Or weather can simply be more extreme than normal. Droughts may be more prolonged. Rainfall can be more torrential. The cold spells can be more pronounced. A blizzard in 1888 actually entombed various Manhattan residents, curtailing 400 lives. In 1993 the so-called 'storm of the century' closed every major airport of the USA's east coast from Maine to Florida. In 1921 Silver Lake, Colorado, received 75.8 inches of snow – over 6 feet! – within 24 hours, a national record as yet unbeaten. Some snow in 1910 slipped from a mountain in Washington state. It then knocked two stranded and stationary trains off their rails into a gorge, killing 96 of their passengers, another record of its kind.

Rain can become flood very easily, when rivers burst their banks. The Arno suddenly flooded in 1966, the worst such inundation that Florence had ever known, and its rampaging waters drowned 39 people. They also submerged 12 square miles of land and damaged a thousand works of art. Canada expects winter, and cold and snow, but a fifth of its population was denied electricity in January 1997. So much ice had collected on transmission pylons that it brought them to their knees.

In John Houghton's book on global warming he includes a table of natural disasters and their fatalities during the 33 years from 1947 to 1980. In order of severity they were:

Tropical cyclones, hurricanes, typhoons	499,000
Earthquakes	450,000
Floods (not associated with cyclones)	194,000
Tornadoes and thunderstorms	29,000
Snowstorms	10,000
Volcanoes	9,000
Heatwaves	7,000
Avalanches	5,000
Landslides	5,000
Tsunamis	5,000

Total human deaths from these catastrophes were therefore over 1.2 million, with an average incidence of 36,700 a year. Their damaging effects on the planet, and on its many forms of life, are impossible to evaluate, but are certainly substantial. Weather in all its forms has occurred on Earth since life began. The extremes, such as those listed above, have also occurred, the catastrophes, the devastations out of all proportion to the normal range of happenings. These things the planet has created for itself, being a multitude of responses to heat from the sun, to heat from within, and to self-generated influences of every kind.

What is not known is the extent to which humans have caused these responses. What is known is that freak weather and natural disasters have recently been most frightening, with 1998 being the very worst of years, according to the International Red Cross. There have been many horrific incidents in the past, but 1998 witnessed a broad range of catastrophe, with 700 'large loss' incidents causing £56 billion in damage. The number of people driven from their homes, and seeking Red Cross aid, rose from 500,000 in 1982 to 5.5 million in 1998. As for deaths, a total of 96% occurred in developing countries. When that terrible year of 1998 receded, and the IRC was able to collate its information, a list showed where much of the disaster had struck during those 12 months.

Month	Location	Type	Damage	Deaths
Jan/Apr	UK	Tornado Flood	£1.5 bn	5
Feb/May	Afghanistan	Earthquake		7,000
Apr	Argentina	Flood	£1.5 bn	20
May	Italy	Flood Mud		119
Jun	India	Cyclone		1,500
Jul	Papua New Guinea	Tsunami		2,000
Aug	China	Flood	£19 bn	4,000
Sep	Caribbean	Hurricane	£6 bn	400
Sep	Sudan	Flood		1,400
Sep	Mexico	Flood		400
Oct	Central America	Hurricane Flood	£3 bn	11,000
Nov	Eastern Europe	Flood	£113 mn	80

With the UK, Italy, Argentina, and eastern Europe being developed areas their combined death total was only 224. The other regions are less developed and 27,700 died.

Some of the events were more general than specific. Droughts, flooding, deforestation, etc. drove 25 million people from their homes, according to the IRC. China's Yangtze flooding affected the lives of 180 million. Flooding overall kills more people than any other form of natural disaster. Flooding in the US causes two billion dollars-worth of damage annually. Severe winter in Russia has helped to devastate a sick economy. In Indonesia, as further consequences of the damaging forest fires, the price of imported rice quadrupled, the local rice crop was unsuccessful, and currency value dropped alarmingly. Attending to such catastrophe, in the Red Cross's opinion, is like raising money for a fire department after a house has caught alight. Steps should be taken now to reduce human exacerbation of natural incidents.

Surmounting, and possibly overwhelming, all of these local occurrences are the extraneous influences, the arrival of meteors and meteorites, the ultimate bolts from the blue. The large kind have come already, and will come again. California is waiting for the big one, the sliding or extending of its main tectonic plates, but the whole planet is also waiting for the far bigger one, the arrival of some most unwelcome hunk of rock. The disruption to be caused in California, when its fault shows – once again – how powerful it can be, will assuredly cause major devastation, but something infinitely worse may be on its way to us, a disturbing possibility forming the next chapter's nub.

Chapter 8

METEORITES

Tunguska – Barringer – Ungava – XF11 – Grove Gilbert – Vredefort – asteroids – AN10 – dendrochronology – comets – detection and deflection

The writer Robert Benchley allegedly did not like sitting in deckchairs 'for fear of meteorites'. He had a point, save that a meteorite heading his way might also pulverise his town, along with him, his deckchair and everything else in that vicinity. A huge one, such as the visitation and catastrophe of 65 million years ago, can even destroy a large proportion of life on Earth by erupting so much material into the atmosphere that solar radiation at ground level is horrendously diminished over several years. Photosynthesis is therefore savagely curtailed, killing first the plants and then the creatures which feed from them. A 'nuclear winter', rightly feared as a consequence of nuclear war if this were ever to occur, would be mankind's mimicry of a meteoric winter, should misfortune of sufficient magnitude happen to come our way.

A word first about words. Meteorology is not the study of meteors, as might seem reasonable, being from the Greek for 'things on high', this phrase arising from the verb to lift. The word could well apply to meteors alone, but it now embraces all of – well – meteorology, the study of the atmosphere with particular reference to its climate and weather. Meteors are the small objects, sometimes merely dust, travelling through space around the sun whose existence becomes conspicuous when they enter Earth's atmosphere, as fireballs if big and as aerolites or shooting stars. They have to approach to within 150 miles of Earth to enter its atmosphere where they become impeded from their great speed, perhaps as much as 160,000 mph, and burn up in consequence. (Astronauts returning from their orbit around Earth also have massive problems with heat, even at their lesser starting speed of 18,000 miles an hour.) Meteoric is the adjective for 'above the Earth's surface' and a meteorite, although sounding smaller than a meteor, is a form of meteor big enough to plunge through Earth's atmosphere and reach ground level. If we did not possess such an atmosphere, and were as naked as the moon, it is believed that our greater size and gravity would have caused us to have been struck 20 times as severely as the pock-marked surface of our natural satellite.

On 12 November 1799 the polymathic traveller, Alexander von Humboldt, witnessed a meteoric storm above Venezuela. On 13 November 1833 residents of eastern North America watched a similar display, with shooting stars more numerous than anyone could count. The suggestion then arose that such appearances were periodic, a theory that was actively promoted by certain individuals and vigorously rejected by others. In 1866 Europe received a celestial exhibition of the same calibre, and the apparent 33-year cycle prompted considerable expectation for something equally magnificent in 1899. The fact that nothing happened, or was observed, provided great scope for mockery from all sorts of quarters, such as journalism, and few individuals were therefore full of expectation in 1933. However, surprising almost everyone, there was a grand meteoric storm over the western US on 17 November 1966.

Everything is now better understood. Those November displays, now known as the Leonids, do occur every year, but only as showers rather than storms. The comet which partners them, 1862 III, is being gradually perturbed by Jupiter's massive gravity. Although Leonid showers have been observed since the ninth century, there will be no more storms after 2164, according to astronomers who have computed that the orientation of their orbit will slowly drift away from Earth. But other swarms of meteors enter Earth's atmosphere each year, notably the Lyrids in the second half of April, the Perseids, usually the most conspicuous, in the first half of August, and the Geminids in mid-December. They leave dust behind in the atmosphere but are not thought to affect the climate, although they might do so indirectly by inducing electrical effects or providing nuclei on which ice crystals form.

By no means are all visitations quite so small and ineffective. 'Falling stones' in China in 1490 killed 10,000 people, possibly victims of some fragmented asteroid encountering Earth. On 30 June 1908 something tremendous hit the vicinity of Tunguska, in south-eastern Siberia. It devastated 700 square miles of trees, but no meteoritic evidence has ever been found in that locality. The impact, equivalent to 20 megatons of TNT, was possibly caused by a portion of a comet or by an asteroid some 300 feet in diameter. (The Hiroshima bomb was only 20 kilotons.) Predominant current opinion is that Tunguska experienced an 'air burst' from a comet rather than an actual impact. Despite the scale of destruction locally, this event is not known to have affected the terrestrial climate. Certainly no other nation then knew that anything untoward had happened, as they definitely did after Krakatau's explosion in 1883.

The Barringer crater in Arizona, Earth's most famous and most visited impact consequence, was undoubtedly caused by something far more devastating than whatever assaulted and flattened Tunguska's trees. The crater's outline is 4,500 feet across, its cavity is 600 feet deep, and the area is marked by a rim some 150 feet above the surrounding land. The meteorite to have caused this tourist attraction is thought to have weighed thousands of tons, to have been 100 feet in diameter, and to have landed about 50,000 years ago. Over 20 tons of meteoritic iron has been collected so far from the vicinity, but no major piece has been found, although anticipated eagerly, deep in the central impact area.

A much bigger crater, named Chubb lake, is at Ungava, Canada. Over two miles in diameter, and with a 400-foot rim, it is even more beautifully symmetrical, and was formed yet more recently, perhaps only 30,000 years ago. Other vast craters, believed to result from meteoritic impact, are at Wabrah, Saudi Arabia, at Henbury, south-west of Alice Springs, Australia, and at Odessa in Texas. In total the Earth's surface possesses about 150 recognised craters (and hundreds more are presumed, some enormous). Australia, so dry inland and so bereft of vegetation, is the world's best place for observing meteoritic impact. It has more craters than anywhere else, with one of them 14 miles across, which is therefore best seen from the air, and another which was formed only 3,500 years ago.

Lake Acraman in South Australia, about 260 miles north-west of Adelaide, is a particularly exciting consequence of a meteoritic impact. Australian lakes do not necessarily possess water, or may only do so from time to time, or merely look as if they might one day, but Lake Acraman created interest by being such a blatant result of a formidable encounter. About 590 million years ago, or so it has been assessed, a rocky object 2.5 miles in diameter travelling at 55,000 miles an hour slammed into this portion of the Australian landscape to form a crater 2.5 miles deep and 25 miles in diameter. The fact of Acraman being an impact site, rather than yet another salt-pan in Australia's dry outback, was only suspected when satellite images showed not only its circularity, but a ring of faults 25 miles across and another ring almost twice as wide. The passage of over half a billion years since the impact means that the crater no longer looks like one from ground level but, once satellites had suggested its ancient history, visiting geologists found a shattered bedrock which provided solid indication of damage by meteorite.

As further evidence of visitations from space many huge meteorites have been found without associated craters, such as a 60-tonner in Namibia, a 14.5-ton block in Oregon, several in Mexico, and also – as it is widely presumed to have come from elsewhere – the famous and revered 'Black Stone' within the Kaaba at Mecca. The best place for finding meteorites is Antarctica. Not only are these alien rocks conspicuous on the snow-white surface, but low temperatures have helped to preserve them from disintegration. A further advantage is that movements of snow and ice have concentrated them in certain areas. Some 16,000 have been found, mainly by American, Japanese and European teams in the past 22 years, which is more than have been found everywhere else on Earth in the past 200 years. The discovered objects are placed in sterile bags, sent in a frozen condition to northern laboratories, and there examined before being stored for later investigation. ALH84001, one particular Antarctic find, is most famous for bearing possible signs of Martian life.

What is of dominant concern, if that is strong enough a phrase, is the thought of something coming from space which is so terrifyingly massive that it would not only destroy everything at the point of impact but damage this planet's climate devastatingly. It might do so sufficiently intensively for all terrestrial life to be in jeopardy (and see Chapter 12 on historical climate change for the event of 535 AD).

This form of concern suddenly accelerated in March 1998 when it was announced that an asteroid, believed to exceed one mile in diameter, was on a possible collision course with Earth and would, in any case, be at its nearest in October 2028. The approaching thing, named 1997 XF11, received publicity via an electronic circular from the Harvard-Smithsonian Center for Astrophysics, at Cambridge, Massachusetts. Reporters picked up this startling information, while other organisations, notably the Jet Propulsion Laboratory at Pasadena in California, promptly made their own calculations concerning this possible encounter of a most disastrous kind. Their subsequent and speedy assessments did not confirm the earlier indication of a 'possible' collision, and the case was closed, within hours of its initiation, when a previously unrecognised image of 1997 XF11 was found on photographic material which had been gathered in 1990. This proved that the asteroid would miss Earth.

The revelation also affirmed that astrophysicists should not rush into print with disturbing forecasts until others have checked the initial calculations. The scientists at Pasadena, although creating relief all round, indicated that this object would approach no nearer than 600,000 miles, or only a little more than twice the modest distance between Earth and its moon. Such a miss can still seem close, and the concern about collision, felt by all of us as well as astrophysicists, has not gone away. Neither should it do so.

David Rabinowitz, of Yale University, and others from Pasadena, summed up the situation in a letter to *Nature* in January 2000. Most small asteroids, with diameters less than 10 kilometres whose orbits approach that of the Earth, have about a 5% chance of hitting the Earth in the next million years. As there are 1,000–2,000 between one and ten kilometres in diameter this means one 'catastrophic collision with the Earth in the next millennium'. Near-Earth objects, or NEOs, which are smaller than one kilometre are not thought capable of creating global havoc by shutting down photosynthesis, but there are many more of them, perhaps 100 times as many with diameters between 100 metres and one kilometre. Apparently there is a 1% chance of Earth being struck by a 300-metre NEO in the next century. Such an impact would be equivalent to 1,000 megatons (of TNT) and could cause 100,000 deaths or, if the affected area was densely crowded, tens of millions. Even if the object landed on the sea it could create tsunamis with devastating effects.

The XF11 publicity, which caused many a red face among astronomers, was avidly discussed at a meeting of the International Astronomical Union (IAU) held in Torino, Italy, during September 1999. This gathering formulated 'guidelines' for astronomers to seek advice, via an IAU-assembled panel, before publicising any potential impact. The Italian meeting also ratified a 'Torino Impact Hazard Scale'. The panel would decide whether a discovered asteroid rated 0 (no chance of an impact) or any number up to 10 (being certain collision causing global devastation). Every decision would be reached within 72 hours, and the panel's analysis of any significant or possible risk would be announced on the IAU website. Astronomers were still free to publicise their discoveries without going through this peer-review procedure but, according to one Swedish astronomer who had attended the conference, 'their statements would have less credibility' than if the individuals making them had followed the recommended policy.

The notion that lunar craters might result from impact and collision is only a century old. The moon's face, so pitted and scarred, was at first thought to have been damaged by volcanoes. The suggestion then arose, from US geologist Grove Gilbert in the 1890s, that forms of meteorite had done the damage. Even in 1949 an editor of *Scientific American* was writing, a little nervously, that 'the most plausible explanation of the craters of the moon appears to be that they were created by the cataclysmic impacts of great meteorites'.

The opposing and volcanic notion only vanished among its strict adherents after the space programme of lunar missions had brought back firm evidence of impact. It was consequently presumed that planet Earth had also been frequently bombarded, even if its atmosphere had served as protection from some of this assault. The attacks were undoubtedly greatest when Earth and moon were young, and when our solar system, together with all its orbiting bodies, was settling into more or less its current form, but many of the moon's craters were formed after life on Earth had embarked upon its exciting evolution. Even during the past few hundred million years, with terrestrial life creeping, crawling, swimming and flying all over the place, there have been large craters formed upon the moon. Therefore it is most probable that Earth, a near neighbour, as well as a larger target with a greater gravitational pull, has been assaulted in similar or even in greater style.

The extinction of the dinosaurs, and of other species in other categories more or less simultaneously, was an intriguing mystery for palaeontologists long before there was any possible explanation. Why should a highly successful reptile group, as well as countless other forms, suddenly vanish from the fossil record 65 million years ago? Various theories were put forward, such as the effect of shifting continents or the proliferation of competitive mammals, before it was suggested that an impact might have been to blame. Calculations concerning the energies involved in crater formation, as with that Arizonan pit, then indicated the massive power of such an event.

A huge crater remnant in South Africa, known as the Vredefort Ring, is 85 miles in diameter, as against the Barringer diameter of less than one mile. The energy needed to create such a massive feature was equivalent to a billion megatons of TNT, according to British physicist Joe Enever, who was first to do the sum. As Hiroshima was destroyed by a detonation equivalent to 20 kilotons the South African explosion was therefore, if this makes any sense to ordinary minds, 50 billion times bigger. (A standard room 50 billion times taller would reach the sun.)

The Vredefort event cannot positively be linked to known species extinction as it occurred two billion years ago, and therefore before most distinctive life began. Microfossils have been found in ancient rocks from Australia and South Africa, proving that terrestrial life, of a sort, was flourishing even by 3.5 billion years ago. This was only 100 million years or so after Earth could possibly have supported life, but these remains were not from forms which can more readily be identified.

Enever's calculations caused people to realise that a subsequent collision, not necessarily so big but more recently in the development of life on Earth, might have been sufficiently devastating in its effects to cause that loss of dinosaurian and other species. The impact would have thrown huge quantities of dust some 30 miles into the atmosphere, and the resultant cloudiness would have greatly diminished sunshine. This reduction in light and warmth would therefore have lessened photo-synthesis and all plant growth. And that lessening of growth would have led to an extinguishing of countless creatures, the herbivores for a shortage of edible food and the carnivores for a shortage of their prey. Game, set and match therefore for the theory, save for the actual location of this impact 65 million years ago at the end of the Cretaceous. To do that amount of damage there would have to be a substantial terrestrial crater – somewhere.

It was already known that a boundary, an 'anomalous clay layer' as it was called, existed between the Cretaceous and the Tertiary, the next geological period. Something major had caused this considerable deposition, and that something could have been an impact. Attention then focused on the Chicxulub area of the Yucatan peninsula in south-eastern Mexico. A crater in that region, partly offshore and partly on land, seemed to satisfy the criteria for such an impact. Its minimum width had been measured as at least 60 miles, this information not only gathered seismically out at sea, where the outline is buried beneath deep sediment, but by observation of certain formations on land, such as sink-holes. The Mexican crater proved to be quite the largest impact remnant to be discovered anywhere on Earth which had been formed since life had developed sufficiently for its effects to be readily found in the fossil record, or in phanerozoic times to use the proper term.

Much discussion then focused on Chicxulub, partly because the forma-tion of massive craters was not properly understood. The large ones do not have the neatness of that Arizona circle. When they are bigger they have a central peak, as can readily be seen with a small telescope in many craters on the moon. When bigger still that central peak becomes a peak ring, and the very largest have several rings, a little like the several ditches guarding prehistoric forts. Careful observation of the craters on the moon has led to these generalities, enabling this work to be duplicated, much more recently, by similar examination of craters on the planet Venus. Corrobor-ation here on Earth of crater formation is difficult because so few large specimens exist, or have been discovered, for such study.

Apart from Vredefort in South Africa, and Sudbury in Australia, which has been much deformed by plate movement, the only other really massive crater found so far is Chicxulub. The total diameter of its impact area, including not only its minimum width but all its rings, has been variously assessed as 110 miles, as 155 miles, and as 185 miles. It is assumed to be a 'peak ring crater', by two of those interpretations, and a 'multi-ring basin' by the third. These discussions will continue, partly to resolve the actual size, but there is general agreement that the Pleistocene 'event' (this little word being much used to describe the biggest cataclysm to occur on Earth since large-scale life began) led not only to the anomalous clay layer but also to the extinction of countless species, not least the dinosaurs.

So what sort of objects are out there in space which might come Earth's way? Foremost are the asteroids, which are presumed to be the fragmented remains of a former planet. They orbit within the solar system between the four small and inner planets (Mercury, Venus, Earth and Mars) and the four giant planets (Jupiter, Saturn, Uranus and Neptune). Not one of these orbiting asteroids is larger than 500 miles in diameter but, as already stated, some 1,000–2,000 of them are more than a kilometre across, a thousand yards or so. About 8,000 have been observed, and there are roughly half a million large enough to appear in photographs collected via the biggest telescopes. So long as these objects stay within their belt there is no problem for us on Earth, but some of them are occasionally perturbed from their orbits by Jupiter's immensely powerful gravity. 1997 XF11 is probably one of those which has been displaced and, with a diameter of 1.2 miles, is neither particularly small nor an outstandingly large specimen of its kind. Even so, and of course disturbingly, it is 3.5 million cubic yards of rock.

Another looming asteroid is 1999 AN10, one more of those greater than 1,000 yards across. At first it was reported that this one would make a close pass on 7 August 2027, with estimates of that closeness varying slightly, but all making it nearer to Earth than half a million miles. According to the earlier estimates it might actually hit our planet, a possibility given the comfortingly long odds of 10 million to one against. These odds were shortened to 500,000 to one in May 1999 when revised calculations, from NASA's Jet Propulsion Laboratory, Pasadena, not only adjusted the likelihood of a miss but changed the possible impact year to 2046. A couple of months later two German amateur astronomers revised that possible impact date to 2076. Armed with up-to-date information these individuals had calculated where it should have been observed during a major sky survey of 1955. Having then spotted it 'on a digitised image' of the earlier records they made their reassessment, boosting Earth's breathing space to three-quarters of a century. Less comfortingly its distance from Earth at the time of its closest path will be a mere 240,000 miles – probably, but with the odds in favour of an actual impact still comfortingly long.

Such impact odds decrease when other and, as yet unexamined, asteroids are brought into consideration. NASA, the US space agency, hopes to identify 90% of these neighbourly asteroids with diameters in excess of 1,000 yards during the next ten years. About 70 new asteroids are discovered each year, any one of which is big enough to upset our climate should it choose to land here. Future generations may not have wars, pestilence or premature death about which to worry, but may look at their calendars, and at the heavens, with far greater fear, knowing when and where some horrific invader is about to strike.

The thought of something so vast hitting Earth inevitably creates concern. Earth's atmosphere serves protectively, and effectively, for the sand-like specks which provide exciting displays when 'shooting stars' are hurtling across the sky; but 3.5 million cubic yards is, as they say, something else. If travelling at speed, which such meteorites would surely be doing, estimated as 100,000 miles an hour or so, the terrestrial atmosphere would only be modestly inhibiting, a bit like an apple's skin, or an apple, or a crate of apples to a speeding bullet. The Vredefort Ring, with its billion megaton detonation, is thought to have been caused by an object six miles in diameter. That is some bullet, at any speed.

Nevertheless smaller impacts are much more likely, with most asteroids being small and with the smallest more likely to be sufficiently perturbed by Jupiter's gravity to come our way. As for bigger arrivals it has been argued that an impact sufficiently large to cause global crop failure is probable every 300,000 years. That sort of statistic is one which human minds find easy to misinterpret, believing that all will be well for that length of time. (Bomber crew's chances in World War Two were 20 to 1 against encountering major mishap. Such crews were allegedly most relaxed during their earliest missions, but became increasingly apprehensive when that total of 20 loomed. Thereafter they could not understand how they had survived, despite the probability of encountering trouble on any particular flight still being 20 to 1 against.)

David Morrison in *Exploring Planetary Worlds* makes death from impact seem almost likely. If a major event were to kill 1 in 4 of the human population, and if it occurred every 300,000 years, the chances for any one of us, and taking the long view, of being killed by impact is a mere 1 in 20,000. Lesser and more frequent impacts, and the lesser traumas they involve, are far more likely and will be far more frequent. Mike Baillie, a dendrochronologist, of Queen's University, Belfast, is an expert on the annual growth patterns of trees. Even the rest of us will have noticed that some tree rings are wider than others, indicating good years for growth at some time in the past. Baillie studied rings for the latest 5,000 years and discovered five worldwide environmental disturbances. He then examined history to see if there were dark ages to parallel these dark times, and encountered some intriguing correlations. His five traumas and their possible historical coincidences were:

2354–2345 BC	Biblical flood
1628–1623 BC	Egyptian disasters at the time of Exodus
1159–1141 BC	Famines at the end of King David's reign
208–204 BC	Chinese famines ending the Ch'in dynasty
536–545 AD	Onset of Dark Ages across Britain

Comets, often believed to be portents of disaster, can also devastate, along with meteorites. Perhaps, and it is a big perhaps, there is reason in this association, with these five traumatic times possibly having been cometary encounters – although the last one is now thought to have been a volcanic upset. Comets are largely formed of ice, rather than rock as with asteroids, but they can still be damaging. A snowball, as we all know, can be painful. A snowball travelling at 100,000 miles an hour would be perfectly capable of doing a great deal of damage, particularly if weighing several tons. No one knows how many comets are out there, all swinging around the sun in their personal orbits. One guess is 5,000, with many of their orbits extremely lengthy. Some, such as Halley's which reappears every 76 years, are short-period orbits. Others, the long-period kind, may be taking millions of years to travel around the sun. Once again the vast bulk of the planet Jupiter may attract and deflect some of them, as it is known to do with asteroids, and then send these objects hurtling towards Earth.

Fact and fiction are happily blended when it comes to movie-making, with truth sometimes submerged to make a better story. The film *Deep Impact* did receive advice from cometary experts, and did accurately portray the kind of assault likely to hit Earth every few hundred thousand years. This could mean next year, or the year after that, or after several hundred thousand years, with lotteries of all kinds casual in their timing as they choose to deliver their form of jackpot. That film, and its successor *Armageddon*, were impeccable in their timing, with 1997 XF11 hitting the headlines a mere 30 years before it was due to hit us most horrendously, as was originally alleged. As further point it is intriguing, if disarming, that the budgets of these two films were greater than the monies currently spent on actually finding and then tracking likely objects heading towards Earth, these visitations possibly confirming what the movies have so forcefully portrayed.

Yet another object, known as 1998 OX4, was found by an Arizonan team before being tracked for two weeks and then lost. A British scientist, one of the specialists in possible asteroid collisions, did not rate highly the chances of finding OX4 again. 'It's like finding a needle,' he said, 'and then throwing it back into the haystack.' Armed with the data which had been collected from Arizona a team at Pisa University calculated that the chances of its collision with Earth in 2046 were one in 10 million, or slightly better than those of winning Britain's national lottery.

John and Mary Gribbin, *New Scientist* correspondents, have argued that 'quite a lot of money' should be available for a 'low probability/high consequence hazard (with) so many people affected by such an event that it is worth taking seriously'. The odds of 1 in 20,000 already mentioned are the same as 'someone in the US being killed in an aeroplane accident, and bigger than the combined risk of death (in the US) from tornadoes, hurricanes, earthquakes, forest fires, and volcanic eruptions'. As David Jewitt has written, in *Nature*'s News and Views: 'The perception of risk from impacts is smaller than for being in a plane crash because planes crash at a steady rate with (relatively) few deaths per event, whereas lethal impacts are rare but kill a lot of people.'

The primary task of identifying potentially hazardous visitors from space is one problem, and it is one which will increase. The rate at which these things are being found has jumped fourfold in the past few years. An American organisation, named Space Guard, has been established to look for potentially damaging arrivals, but has received only modest funding. Its primary intent has been to look for 'Earth crossers', those asteroids whose orbits approach Earth's own revolutions. In December 1999, as requested by Britain's science minister, the UK established 'a task force' to assess the risk and possible damage of a meteorite hitting Earth. This four-man team has been asked to prepare its initial report on the problem by May 2000. It is generally believed by professional astronomers that it will be amateurs, with no funding whatsoever save from their own pockets, who will be the ones to detect and locate Earth-threatening invaders from space.

There is also a secondary problem: how can these rocks be deflected from their paths once it has been established that they are coming our way? The Gribbins argue that it took the US less than ten years to put a man on the moon after John F. Kennedy had 'fired the starting gun'. Now, with space technology 40 years more advanced, and with the resources of the entire world available to divert and prevent an impending global catastrophe, XF11 'could have been deflected', had such a course been thought advisable. David Morrison, in *Exploring Planetary Worlds*, suggests that $500 million a year should become immediately available to detect and then prevent cosmic impacts. John Maddox, emeritus editor of *Nature*, summed up his views on tasks for the 21st century in January 2000: 'If the objective is survival of the human race, we would surely be spending more on carbon dioxide abatement, less on biodiversity and a lot more than we do at present on avoiding asteroid impacts with the Earth.'

Procedures for deflection were recently discussed at the Los Alamos laboratories in New Mexico, where the Manhattan Project developed the atomic bomb. These included: massive mirrors in orbit which would scorch and distintegrate the offending object; the attachment of sails to an invader which would make use of solar wind for diversion; the fixing of small rockets which would act in similar fashion; the placing of great quantities of dust in the invader's path; and, somewhat inevitably, the detonation of nuclear devices. Edward Teller, generally credited with siring the hydrogen bomb, and who used to advocate all manner of 'peaceful' uses of nuclear weaponry, was – a touch surprisingly – against this last idea.

An ultimate solution, also mentioned at the conference, was to realise the inevitability of an impact, to foresee the destruction of all life on Earth, and to evacuate a quorum of individuals to Mars, of the type already encountered by James Bond in *Moonraker*. They would then live within an artificial environment, and wait until Earth had once again become habitable before returning and, as Genesis originally instructed, being fruitful, and multiplying, and replenishing the Earth. Such a return would be equivalent to the Krakatau story where all terrestrial life was obliterated before, quite swiftly, making a comeback.

Remembering what happened at Tunguska, in historical terms just the other day, and wondering what would have happened had 700 square miles of urban development been flattened instead of a multitude of forest trees, such an expenditure could be quite a bargain. Manhattan is 31 square miles and London about 100. Robert Benchley would assuredly have donated funds aimed at making deckchair relaxation less troubling as an activity, or might have abandoned this procedure absolutely if he had learned about XF11 and the rest, confirming his suspicions that a much-loved planet is an undoubted target for wayward things from space – from time to time, as and when they choose.

Chapter 9

BIOLOGICAL RESPONSE

*Palolo worm – breeding cycle – suspended implantation – daylength – hormones –
migration – lemmings and monarchs – hibernation – torpor – aestivation*

When talk of global warming is in the air – or by the shrinking glaciers, the
extending deserts, the changing ozone layers – it is often regarded as a
problem only for humans. Human dwelling places may be submerged.
Their food crops may not grow. And, as individuals, *Homo sapiens* may
suffer from ultraviolet radiation, along with all the other changes. Human
beings find it easy to be myopic, viewing man-made changes solely for
their effects upon man-made livelihoods, much as the sins of the fathers are
said to fall specifically upon their children. But we share this planet with
every other form of life, and all those forms will also experience any
changes which we alone have wrought.

It is additionally easy for humans to disregard the fact that other life
forms are as dependent upon the calendar as we are, however much we,
via our ancestors, have invented weeks and months, along with leap years
and days of rest, for our convenience. What we have not invented are the
days themselves or the annual seasons which partner Earth's peregrina-
tions around the sun. Ever since life began on our planet, and even before,
there have been days and nights. There have also been years, with their
remorseless cycles of dry times and wet times, of warmer days and colder
days, of longer nights and shorter nights. Humankind did not create the
astronomical situation, the solstices and equinoxes, the perihelion and
aphelion, or the manner in which this planet is heated by the sun. It merely
took note of these phenomena, just as living things are also aware, in their
own fashion, of the progress of each year.

Palolo worms are a striking case in point, reminding their neighbouring
humans of the actual date as keenly as any timepiece. This lowly
polychaete, *Eunice viridis*, lives in rocky crevices on the shorelines of Fiji and
Samoa. For most of each year the species remains hidden, seemingly

unaware of anything. Then, at dawn, on both the very day and the day before the October and November moons enter their last quarter, this animal spawns. Even more incredibly it is therefore incorporating the lunar cycle with the annual cycle into its once-a-year astonishment, a feat which humans have failed to achieve with months.

The palolo may be a simple form of worm but its awareness of the passing year, and how far each year has progressed, is plainly more embedded in its system than most humans are able to achieve without benefit of calendars. The animal keeps its sense of time, despite being warmly bathed by tropical sea-water all year round, and receiving steadfast sunshine throughout the year, living as it does not far from the equator. Its extraordinary ability remains a mystery, however fascinating. (Unkind zoologists have likened this remarkable creature to the British Royal Society which also reproduces at that time each year, when it too surfaces to make good its numbers via the creation of more fellows.)

Life on Earth is remarkably adept at knowing the season of the year, whether or not the year is being, as meteorologists phrase it, unseasonable. Take the case of an ordinary non-migratory species of bird living in the temperate zone. In January it starts defending its territory. In February it pairs up with one of its own kind. In March, when both sets of gonads have ripened, it mates. In April it builds a nest and lays eggs. In May its young are hatched. In June, with insects abundant and the days at their longest, it busily feeds the growing nestlings. In July these youngsters fly possessing, as yet, undeveloped gonads, which are therefore equivalent to the now regressed gonads of their parents. August is the time for parting, when parents and youngsters separate. In late autumn all gonads begin to develop, once again for the parents, first time for the young. By each year's end, when the days are darkest and insect life is at its minimum, these organs are then growing rapidly, knowing in advance, as it were, of bountiful times to come. This process happens every year. It is as natural as night following day. It is also mind-blowing in its complexity and precision.

Compromise is also pertinent every inch of the way. Insect food, although important for many species, is not the sole criterion. There is also the length of day, with daylight and feeding possibilities almost unending during summer's peak. Another consideration is location and its permanence. Is migration a better option? To travel further north in the northern hemisphere is to encounter longer days after 21 March. There will be more time for feeding the young, certainly. The temperature is colder than further south, probably. There will be fewer resident

predators, probably. There may be less food, notably of insects. All springtime changes will be happening later in the year, most definitely.
Therefore everything has to be more rushed. Will there be time for a
second brood? Probably not. In short, it can be better to save the migration
energy, to remain more or less in the location of one's hatching, and then
make do with all its disadvantages along with its varied virtues.

In springtime, so it is said, a young man's fancy lightly turns to thoughts
of love, but there is little point it doing so, biologically, unless gonads are
already functional. Timing, save for humans and apes, who can mate all
year round, is a matter of anticipation, of knowing precisely how far a year
has progressed, of making preparations in good time, much like Christmas
card manufacturers who are possibly busiest in July. Most humans are
pathetically inept about the length of passing days, only noticing and
saying that they 'are drawing out, or in' when they have long been doing
so.

Deer, on the other hand, are generally in rut within one week after the
summer solstice. Birds tend to get their timing right despite abnormally
cold winters, delayed plant growth and miserable weather. They take
more note of changing day length, the guaranteed and entirely reliable
marker point of each year, with that change either shortening or
lengthening. Increasing day length is the stimulant for ferrets, hedgehogs
and raccoons, for example. Decreasing day length is, conversely, the
stimulant for sheep and goats.

There is also the problem of gestation. Mating should be timed so that
progeny are born during the most beneficial season. Sexual activity cannot
take place in the spring if the offspring are to be born in the spring, save for
those animals whose pregnancies last for 12 months, such as (more or less)
the horse, the llama, the badger, zebra and wildebeest. There are also no
seasonal complications for those other species, such as small rodents,
whose gestation times are extremely short. With only three weeks between
copulation and birth for mice, and only another six weeks between birth
and parenthood, several mouse generations can be produced during the
warmer, food-rich time of year. Some speedy birds can lay and rear three
clutches in a single summer, although the third clutch will probably fail to
survive if too delayed and the chill of autumn arrives too swiftly.

A trick concerning timing and pregnancy involves delayed implantation
or, a term more favoured by biologists, embryonic diapause. These terms
define a hesitation in pregnancy development, an arrangement favoured
by many marsupials, by seals, some deer, and various mustelids akin to
stoats and weasels. It is a stop-start procedure. Fertilisation takes place

causing development to begin, but then it halts. Much later it starts up again, and will then proceed at a normal pace as if nothing has occurred, much like traffic when finally released from a jam.

The Atlantic grey seal is one example. From copulation to delivery is, on average, 351 days. Therefore mating and birth can take place during the same season of the year when the animals have hauled themselves from the water on to their mating/breeding grounds. After each successful copulation, and for six days thereafter, embryonic development takes place normally. Then virtually nothing happens by way of cell division for the next 100 days. After this three-month hesitation the foetus then begins to develop once again, and the young are born after a pregnancy of 11.5 months rather than the 8 months which might reasonably be expected if there were not such a hiccough in development. Two weeks after each birth there is another bout of intercourse, which initiates the next round of delayed gestation. Stopping the system, temporarily, was plainly easier for evolution to achieve than slowing down the whole process. Even a human being's leisurely development, where all emphasis is on slow maturation, takes 12 gestation weeks fewer than the seal.

The phenomenon of arrested development was first described in roe deer. These ungulates mate in late summer and produce their young in May. The central portion of summer can therefore be wholly dedicated to the care and protection of offspring. Roe deer foetuses need about five months to complete their development, but a birth during winter's coldest days would be unsatisfactory and possibly lethal. Therefore the three-month delay, the hiccough, the embryonic diapause, causes roe deer young to arrive more sensibly in May.

Certain bats achieve the same result via a different route. These (usually) small creatures might be expected to have short gestation times, and indeed they do, but there is a hesitation between copulation and birth, with mating occurring in the autumn and birth taking place during the much later and insect-rich days of summer. With bats it is not an interrupted pregnancy but a prevented fertilisation which causes the delay. There are no eggs for the autumnal sperm to encounter until the following April when, after half a year of waiting, the sperm are finally able to fulfil their purpose. Development then proceeds apace, with the young born in June or July when food is conveniently at its peak.

Latitude and day length are not of major importance to human beings, provided their food crops can grow sufficiently or their prey are reasonably abundant. For birds busily raising their young the number of hours per day available for feeding can be critical. Comparison has been made between various pairs of American robins, such as those living in Alaska and others in Ohio. The southern birds were able to achieve 96 feeding visits per day, on average, while the Alaskans, with far greater day length, could manage 137. The northern young could therefore leave their nest after 9 days as against 13 for the fledglings in Ohio. The nest-time period, one of considerable vulnerability for the youngsters, was therefore reduced in Alaska by 30%. As an extreme generalisation birds do not nest and bring up their young whenever or wherever daylight lasts for less than 11 hours.

The undoubted benefits of living further north have to be set against the demerits of a later spring and a shorter summertime. The family *Fulicariae* of coots and moorhens makes this point on its own, ranging as it does across the latitudes. The more northerly pairs of birds perform their breeding later in the year. In general terms each coot egg-laying season starts 20 to 30 days more tardily for every 10 degrees of latitude north from the equator. There has to be a limit with this delaying tactic, and the coot family does not breed north of 60 degrees N. There must, of course, be sufficient time to raise a family.

With altering day length acting, almost universally, as trigger for the sexual cycle in birds and mammals it is the arrival of warmth which is generally more important for those animals needing outside warmth for their activity. Some reptiles use light as stimulant, but most do not. As for amphibia they respond only to warmth, save for the species which use rainfall as a starting gun. Fishes, somewhat inevitably, do not take note of rain, but respond to warmth and maybe light as well.

As for invertebrates it might be expected that heat would be the prime criterion, but the palolo worm's behaviour must not be forgotten. It never fails, and its breeding activity responds only to the last quarter of the October and November moons. What is so different about those moons to all the other last quarters of all the other moons each year? And are invertebrates, in general, more alert to the changing seasons than might be suspected from a casual observation of their different ways? It is so easy to be wrongly arrogant about smaller creatures, calling them primitive or simple or merely lower forms of life. Their complexities are still far from being unravelled and, in time, they may be seen as much less lowly than originally and currently believed.

The standard temperate situation, with animals breeding in springtime, with trees then coming into leaf, and with the warmer days of summer better both for day length and for food, can make those of us who live in the temperate world assume this pattern is typical of all life; but not so. In the tropics tree species tend to flower in their own time, uniformly as a species but not as a forest. As for triggers, with changing day length near the equator less detectable, cloudiness may be more important, or ultra-violet intensity, or rain, or even the fact that other trees are already flowering. Tropical plant species relying upon day length have to be extremely sensitive to the modest differences from one day to the next, and so they are, with rice and cowpeas known to detect alterations of 10–20 minutes. Latitude is all important. At Entebbe, for example, the longest day in any year is 12 hours, 50 minutes; so too is the shortest if twilight is included. At Hawaii the annual variation is 2 hours, 27 minutes; at Philadelphia it is 5 hours, 39 minutes; at the Shetlands it is 14 hours, 37 minutes; and in North Alaska it is 24 hours.

Within the tropics there is certainly not the on-off and straightforward simplicity of the temperate world. Some birds living nearer the equator, such as Madagascar's fantail warbler, breed throughout the year, but even this species has a breeding season in other parts of its range. *Miniopterus australis,* a bat which spends all daylight hours in a dark cave, and is therefore largely immune from changing day length, is somehow aware of the changing year as its females only become pregnant at the beginning of September.

Temperate dwellers consider it entirely reasonable that swallows arrive in Europe during springtime to start their breeding process. Not only do the birds then encounter lengthening days but they meet abundant food, good niches, and a period of increasing warmth, although less warm than the African continent they have left behind. When breeding has been completed, and days are shortening with food becoming less abundant, it seems most reasonable to temperate humans that the birds fly south once more. On this journey, to southern Africa for the swallows from Britain, the birds soon encounter lengthening days, abundant food, good niches for survival, increasing warmth, and circumstances entirely favourable for breeding, but they do not breed. Their internal sexual rhythm can only be triggered once each year.

The precision of breeding cycles, with mating either welcome and worthwhile or unappealing and valueless, must also have been a cause of speciation, the separation of one species into two. If one group comes on heat/is fertile a couple of weeks before another that distinction will then be perpetuated. The animals may meet, but they will never mate. Therefore any mutations affecting either group will not be shared by the other, and the two groups will become more and more distinct, even though timing of the breeding cycle was, at first, the only difference.

*

Underlying all this varied information has been the production of hormones, named after the Greek for 'excite'. There is not some magic wand causing oestrus to begin, gonads to enlarge and eggs to ripen; everything of this nature occurs under hormonal control. The principal hormone producer, and master key of the reproductive cycle, is the pituitary gland. This diminutive object, generally egg-shaped in vertebrates but far smaller than (chicken) egg-sized, is attached to the base of the brain by a short stalk. It is formed in two distinct parts, of which the anterior lobe, and its secreted hormones, is of most importance to the reproductive cycle.

Endocrinologists may, and do, revel in the complexity of the hormonal cycles, as chemicals are created to stimulate or depress the production of other chemicals, but the actions and interactions can be mind-numbing for other individuals. As example, the fact of a mammalian egg being prepared within an ovary for a possible encounter with sperm can seem an elementary happening. It is, after all, a fundamental event, but the hormonal involvement can confuse. In broad terms, and to summarise what happens:

> The anterior pituitary secretes a follicle stimulating hormone (FSH). This causes the egg follicle to secrete oestradiol. The oestradiol acts on the anterior pituitary to decrease the amount of FSH. When the level of FSH falls sufficiently a luteinizing hormone (LH) will be secreted by the pituitary which also affects the ovaries. Yet another pituitary hormone, luteotrophin (LTH), then acts upon the corpus luteum – tissue remaining at the site of the egg follicle – to produce progesterone, the hormone which maintains a pregnancy. As the corpus luteum develops it steadily depresses the production of LH and LTH. If, on the other hand, a pregnancy has not been achieved, the anterior pituitary will secrete FSH once more and another follicle will be stimulated.

No wonder it is easier to fall back on some direct statement, such as 'eggs are produced by the ovary in the spring', but the remark should not be left as if complete in itself. What is it about spring which causes the pituitary gland to begin secreting? Or, perhaps more accurately, what is it about changing day length around the time of the winter solstice which causes eggs to be ready for fertilisation in the following spring? And, after remembering that two sexes are involved, what is it on the male side which causes synchrony, so that each male testis is at an appropriate stage of development to match the ovary's development when that spring arrives?

Most vertebrate máles remain sexually potent, whether or not their females are undergoing cycles, but some have cycles of their own, such as deer with their annual rut. And some have their sexual cycles triggered by nothing seasonal, being entirely influenced by their sexual opposites initiating the seasonally induced courtship process. Once again such a statement implies simplicity, but sexual appeal and repulsion are extremely complex interactions. In the spring a young man's fancy does indeed turn to thoughts of love, an occurrence no less remarkable than every other biological consequence of Earth's steady circumnavigation of the sun. That annual cycle creates an astonishment of other cycles, each being the effect of an astronomical happening lying at the root of all our seasons.

Migration merits a section to itself, being so remarkable and also planetary in its scope and range. At first sight it seems extremely sensible: travel when need be to wherever the conditions are more favourable, much like the wealthy English who used to voyage to Egypt in February, to Rome in March, to Paris in the spring, and then to London for 'the season'. On the other hand the energy expenditure of migration is formidable, the hazards en route are considerable, and no sooner are destinations achieved than it is almost time to travel back again. Even the wealthy English dowagers did not journey further than they considered absolutely necessary, and certainly did not mimic British-born swallows in travelling from over 50 degrees North to 30 degrees South every autumn, only to hurtle back again a few months later. That straight line distance of 11,000 miles every year is an amazing undertaking.

The Arctic tern is the greatest migrant of them all in terms of distance. Many of this species breed north of the Arctic circle before travelling to the Antarctic as soon as the frenetic business of northern hatching has been achieved. Each half-year journey is therefore in excess of 9,000 miles, with the migrations alone, apart from feeding diversions, demanding an average of 50 miles a day every single year. The principal advantage for these birds is that they always experience daylight for at least 12 hours in every 24, and often very much more, with even 24 hours on occasion out of the 24.

Greater shearwaters, *Puffinus gravis*, are similar, with the starting point and final destination being, for some of their number, the volcanic speck in the south Atlantic of Tristan da Cunha. These Tristan birds fly north after breeding from January to March before congregating off Newfoundland in early June, only to disperse for various portions of land in the north Atlantic. Then it is back again, and even less directly, to that island retreat. Perhaps it is more reasonable to consider their home as the entire Atlantic Ocean because they treat it much as domesticated ducks regard the village pond.

Most birds migrate a little, even if for far shorter distances than those of terns and shearwaters, and some undertake astonishing flights. Certain geese fly over the Himalayas, and have been recorded higher than Everest where oxygen is one-third the sea-level density. Even curlews and cranes have been observed at 20,000 feet. Migratory speed can also be remarkable, as proven by ringed birds on recapture. A blue-winged teal once flew 3,300 miles in 27 days, a sandpiper 2,300 miles in 20 days, and a lesser yellow-legs, an American wader, 1,930 miles in 6 days, these three birds therefore averaging 122, 115 and 321 miles per day respectively. Wilson's storm petrel, which usually nests south of South America, is a small and slowish bird, but then spends each northern summer in the north Atlantic. Petrels in general, along with albatrosses, treat the whole world as their oyster, much like today's well-heeled tourists exploiting every hemisphere, north, south, east, and west, with similar disdain.

Most marine fish migrate, from feeding grounds to spawning grounds. Many river fish, such as salmon, start life in a river, feed in the ocean and return to that same river to breed. Whales also voyage from good sources of food to good, it is presumed, breeding sites. Turtles are tremendous migrators, going somewhere else, often who knows where, before returning like nostalgic humans retiring after a lifetime of travel to the spots where they were born. Chitons, small invertebrates rigidly attached to rocks, can seem the most static of creatures, about as venturesome as the granite on which they sit, but Bermudan chitons have been observed travelling a yard each day before inching back to their favoured dwelling sites. And, as fleas have other fleas on their backs to bite them, such molluscs often have far smaller molluscs on their protective plates, these also migrating each day upon their chosen carapace.

Two particular migratory creatures deserve special emphasis, their stories being so extraordinary. One is a vole-like rodent and the other a butterfly. *Lemmus lemmus*, the Norwegian lemming, is famed in legend for pouring from its hilltop homes, and making victims of humans by consuming crops remorselessly, by fouling water, and by hurtling after their leader with a mass suicidal passion until every single life is extinguished by swimming out to sea. Like all good stories little of that is true. Instead, about every fourth year, this species is subjected to outbursts. These are never true migrations because a return journey is not part of the cycle. There is no leader, and the animals themselves frequently become the victims when dogs and every other form of predator take advantage of a sudden bounty of minute meat, average lemming weight being 2.5 ounces or one-tenth the weight of a guinea pig. The scurrying forms, only travelling by night as is their nocturnal manner, try to evade rivers and lakes if possible, but nevertheless do die in tremendous numbers, although there is not a total loss of life.

The lemmings are not alone in this form of perversity, only the most dramatic. Such one-way migration is difficult to comprehend. With *Lemmus* it is not caused by a lack of food, or by an outbreak of disease, but happens when the resident population has reached a certain density. From an evolutionary viewpoint it is even harder to understand. Who goes, who stays, and why? Those who remain are the ones who provide the genetic material for the next assortment of goers and stayers, and yet they have not gone themselves, despite the probable urge as tens of thousands of their fellows suddenly depart. This Gadarene behaviour has been properly recorded for almost 500 years, and has not diminished. It has not been selected against in the evolutionary process, as might have been expected, and it is still hard to fathom.

The problem also makes us question other migrations which seem to make sense, and which therefore seem, from our viewpoint, to be advantageous to the species. Are they too driven by a lemming-like impulse which makes no sense, save that we have found a happy interpretation? Would it actually be better for Arctic terns either to ride out the northern winter or merely travel much less far? We tend to say that nature knows best, but we say it more quietly when we cannot discern any reason in some apparent madness.

Monarch butterflies, *Danaus plexippus*, pose further questions with their extraordinary migration. This insect's neural system cannot be equated with a bird or mammal brain, and yet the annual trip from north to south and back again is no less extraordinary than many vertebrate migrations. The diminutive lemmings and starling-sized storm petrels are considered minute for their tremendous travellings, but the Monarchs are only butterflies, pathetic lightweights by comparison. And yet the flight time for a measured 1,870-mile journey lasting four months from Toronto to Mexico averaged 15 miles a day. Whether the youngsters hatch either in Canada or the northern US they do not start as a horde, which is the lemming experience as well as that of countless other migrants, but each sets off independently, the urge to travel being an instinct embedded in every single set of genes.

Each winter these insects are in a dormant state, hanging from trees bordering the Gulf of Mexico or near Florida's and California's shorelines. With the arrival of April and of greater warmth they become active, with each aroused individual soon determinedly heading northwards. They fan out as they go, and will eventually lay their eggs on *Asclepias*, the milkweed plant, which is sufficiently abundant around the central belt of North America. The more southerly eggs are first to hatch, doing so towards the end of June. These progeny then have time not only to become sexually mature but to lay eggs of their own. Canadian eggs, laid later and hatched later, usually only produce a single crop of young.

All the surviving butterfly progeny, whether the solitary generation to be hatched that year, or a second generation from that summer's earlier hatchings, then head south. They leave behind them all those parents who flew north for the summer, for every one of these older butterflies dies at the end of the summer season. The newcomers therefore have no guidance, save for the compulsion within them to fly south and then seek out coastal areas, much as all their parents did one year earlier. Each butterfly sets off independently and, although every single one heads south, they can end up in a wide variety of locations. A set of eight, hatched near Toronto and closely monitored, ended up in New Jersey, South Carolina, Alabama, Louisiana, Texas and Mexico, with five of the eight in coastal areas. When cooler weather arrived in all these different areas the several insects being examined each hung from trees and became dormant, precisely as their ancestors had done one year earlier.

This form of suspended animation which the butterflies experience cannot be called hibernation, as that phenomenon does not apply to cold-blooded creatures whose internal heat falls in tune with the ambient temperature. Hibernation is quite a different happening, and another form of adaptation to winter's chill.

Four kinds of response are possible when confronted by uncongenial circumstances. The first is to say Goodbye and emigrate. The second is to leave, but then return when conditions are better; in short, to migrate. The third is to remain stoically in place, possibly to die. The fourth is to adapt and make an unpleasant situation less exacting.

Hibernation lies in this final category, being a form of abnegation, a withdrawal from the struggle of existence when conditions are unfavourable. It is not a kind of sleep. Neither is it the sort of torpor or dormancy found among certain mammals (like the brown bear whose body temperature only drops a few degrees), among birds (like the swift), among fishes (like mackerel and basking shark) and among insects (like queen wasps and those Monarchs). Instead it is a controlled drop in temperature, and not merely a lowering of metabolism. It used to be thought that mackerel and basking sharks migrated elsewhere during the winter because they were only seen, and could only be caught, during the summer months. Much later it was presumed that they spend the non-plankton season on the seabed in a reduced state of activity, but this is not considered to be hibernation.

As for true and proper hibernation, some animals, which normally maintain a high and constant body heat, abandon their warm-blooded (homeothermic) state and permit that heat to fall close to the environmental level, a drop which is never lower than 0 degrees Centigrade, save in some bats, and a few exceptional other mammals, such as Arctic squirrels. All hibernators will, when conditions become more favourable, then permit their internal heat to rise to its normal level, even though the outside temperature of spring may still be very much colder.

Human beings can do nothing of the sort, partly for being so large with even the biggest hibernators being less than one-twentieth of the adult human size. A small creature, with all of its body close to the exterior, can warm or cool very rapidly. Large creatures would not be able to do so, a drawback preventing hibernation, but humans can on occasion be cooler than their normal temperature. Old people are particularly prone to becoming chilled during cold conditions, often to 95 degrees Fahrenheit, and therefore to 3–4 degrees below normal. In severe situations they can cool to 90 degrees, or to temperatures even colder, but this chilling process is insidious. From 98 F. to 90 F. there is a feeling of cold, and shivering occurs. From 90 degrees to 75 the unfortunate individual is depressed, but is usually, and disconcertingly, uncomplaining about the lack of warmth. During this time of reduced body heat the pulse will slow down, blood pressure will be reduced, and shivering will stop at 85 degrees or so. Below 75 F. the path is downhill all the way, with survival rare. Mankind is definitely not a hibernator.

Conversely the hedgehog, *Erinaceus europaeus*, gains from being one. Its spines, good for protection in general, are poor insulators, and nothing like so effective as fur. Moreover the animal's traditional invertebrate diet of insects, worms and slugs disappears in winter, but it is not this lack of food which acts as spur. Instead it is cold weather, perhaps in December, which serves as trigger for these animals to become comatose and to start their hibernation. From a normal heat of 93 degrees F. their bodily warmth falls almost in keeping with the outside temperature. Their heartbeat drops from a normal 128–210 per minute to 2–12. Their metabolism is much reduced, falling to one hundredth of its standard quantity, and their internal temperatures lessen as if the animals themselves are blocks of wood; but, with this species, body heat never falls, except in death, lower than 43 degrees F. (This is the same temperature, coincidentally, below which most plants no longer grow.)

For hedgehogs, if the wintry conditions are being particularly harsh, they may even wake up, as if deliberately, from their somnolence, and generate some heat. In any case their torpid state is not continuous. Every 8–12 days the animals arouse themselves. This causes their metabolism to increase, their heartbeat and respiration to go up, and some warmth to be regained before their temperatures are allowed to fall once more as the previous somnolence is reinstated. There are parallels here with human sleep which, however seemingly log-like to the contented sleeper, is repeatedly interspersed with movement. Hedgehogs eat nothing during their hibernation. Therefore weight loss is inevitable, notably at the start.

Who else hibernates? Several hundred species from six mammalian orders do so, and all are small. The real lightweights are the 20-gram Californian pocket mice. Relative giants are the three-kilogram marmots. Despite entrenched human opinion to the contrary the grey squirrel of America, and of Europe since its introduction, does not do so. Neither do any of the tree squirrels, but ground squirrels living in cold places, such as the Arctic species known as sik-sik, are true hibernators, consuming only one-eightieth of the energy they would use if active at that time.

This most northerly hibernator is 50% fat when initiating its eight-month absence from the world each September. The deposited nourishment has to be of the right polyunsaturated fat if it is to be accessible as food when its possessor has become cooled in wintertime. Saturated fats only stay liquid at normal body temperatures, but no mammals can create polyunsaturated fats, any more than they can create most vitamins. Therefore their summertime diet must contain them, such as the lineolic acid of many nuts and seeds. The squirrel's permafrost burrow may drop to minus 25 degrees C., and the animal itself may lower its temperature to minus three degrees. This supercooling does pay dividends. Every ten-degree drop in body heat halves the rate at which energy is consumed. Arctic squirrels can also sleep lengthily in summer whenever there is drought and food is scarce. The deep 'sleep' of hibernation is quite different, with EEG recordings from an electroencephalogram showing, essentially, a series of flat lines. Humans would be considered 'brain dead' if possessing such even tracings.

Marmots are also proper hibernators, waking up every three to four weeks to defaecate and urinate before settling down again. All true hibernators, even that Arctic squirrel, perform this waking up routine for perhaps half a day every few weeks, even though the procedure is extremely costly in its use of energy. The squirrels consume 80% of their fat during this warming process, when body heat may rise to plus three degrees, and the animals neither eat nor leave their burrows during this disturbance.

Dormice certainly hibernate, their name from the French dormeur/ dormeuse for sluggard or sleeper. (They are a form of rodent, but are certainly not mice.) In Europe they hibernate from October to April, and tend to mate the moment they surface from their resting state. Much to general disbelief, but a well-known fact to the local Hopi Indians who called it holchko, the sleeping one, there is a hibernating bird. *Phalaeonoptilus nuttalli* is a poor-will nightjar and, in southern California, a specimen was found in a rocky crevice for three consecutive winters. Its measured temperature proved to be 18 degrees C. Birds customarily have higher temperatures than mammals, with the normal heat of chickens being 41.6 degrees, an internal warmth likely to kill most humans.

Certain bats are the most extreme hibernators. Their body temperatures may even drop below 0 degrees C., the freezing point of water. According to David Macdonald's *Encyclopaedia of Mammals* the lowest temperature ever recorded is minus five degrees C. for *Lasiurus borealis*, the red bat. This splendid volume of information also reports that 'old adult females are first to begin hibernation, followed in succession by adult males and juveniles'. (There are parallels here with human households in the night-time order of retiring.) Before their hibernation the bats of temperate zones accumulate considerable food reserves, mostly as fat deposits which can be one-third of the total body mass. The lower the bats can reduce their temperature during hibernation the longer these reserves will last.

Pseudo-hibernators are more widespread than the true examples, such as those truly hibernating bats. Torpor, as a state, is difficult to define, being a kind of halfway stage. Animals in torpor have lower temperatures than normal, but these are nothing like so low as with the genuine hibernators. Swifts can experience torpor, and so can humming birds. Both groups have a tremendous demand for food, and this is lessened if the birds, temporarily, reduce their metabolism. Some mammals use the same trick, even daily, much as siestas are a way of life for some. Torpor is the poor relation of true hibernation, also beneficial but nothing like so intriguing, or so improbable, as the ability of a few warm-blooded creatures to lower their all-important body heat so dramatically.

The need for torpor can even be investigated arithmetically. This has been done with the basking shark, second largest of the 370 species in the shark family after the whale shark. This fish, commonly over 30 feet long and even measured up to 45 feet, maintains its tremendous bulk by filtering plankton. To acquire this food it travels through the water, with mouth agape, at some two miles an hour. The power needed to propel a 22-ft shark with a 28.5-inch mouth presenting a frontal area of 4.3 square feet has been calculated as one-third of a horsepower, equivalent to 212 kilocalories an hour. As the shark's propulsion system is undoubtedly imperfect in design it has been further assessed that the real hourly output is probably nearer 265 kilocalories. The procedure, via digestion, of converting all the food energy absorbed into the energy required, namely of intake to output, is also far from perfect. Hence a further estimate that actual food requirement is likely to be nearer 663 kilocalories per hour than 265. A final assumption, necessary for the arithmetic, is that a 22-ft shark swimming at two mph would be able to filter 16,000 cubic feet of water an hour.

In wintertime around Britain it is known that such a quantity of water contains only about 2.75 pounds of plankton, a weight of food equivalent to 410 kilocalories. To expend 663 kilocalories in order to gain 410 plainly does not make sense. Hence a presumption that basking sharks either migrate to warmer and plankton-richer waters every autumn or, somehow, cut down on their requirements. The suggestion that they rested during wintertime on the sea bottom was therefore a satisfactory answer for the, otherwise, unsatisfactory arithmetic.

Unfortunately there is no proof of this relative immobility, and no indication that such recumbency is either attacked by scavengers or colonised by plants or other creatures taking advantage of a stationary object. (The sea bottom is said by many to be less well known than the surface of the moon.) It is now suspected that some of the arithmetic's basic information is wrong, such as the food value of plankton, and each shark's metabolism may merely tick over quietly during wintertime. The animals would therefore acquire their necessary nutrition from the squalene oil stored in their livers, a property of all sharks, and would spend very little energy in pointless movement and the unrewarding procedure of gathering food when food is not there to be gathered in sufficient quantity.

It is also believed, by some, that the sharks lose their gill-rakers in wintertime, this plankton-gathering equipment being no longer necessary and therefore a waste of energy to maintain. A little like trees shedding their leaves in autumn, and then regrowing them in spring, a profit and loss account is involved. Is it more worthwhile either to keep a less profitable organ or to discard it and fashion it anew when better times arrive? The subject of bio-energetics involves a critical examination of effort-to-yield, of the overall worthwhileness of any activity. With basking sharks there is, at present, insufficient information about their lifestyle to assess the effort it involves.

Such arithmetic can be applied to every aspect of seasonal behaviour. Is it more profitable to fly 6,000 miles or to stay put? The migrant's energy expenditure is colossal, but so can be the disadvantages of remaining in one place. Is it better to hibernate, and consume stored fat, or carry on and hope to find sufficient food? The hedgehog gives up living, so far as it is possible to do so. The squirrel remains active, often even producing progeny in wintertime. Is it also better, as some hummingbirds do, to switch off and become dormant from time to time, rather than persevere with the near-perpetual search for nectar? And is it also profitable, as with some particular hummingbirds, to fly over water for 24 hours between the Americas, having stored sufficient nourishment for the journey? What have they gained by arriving in another habitat, the journey having been so hazardous and so exhausting? Profit and loss accounts are almost as varied as there are species.

A further compromise is aestivation, a state of torpor during summer's heat equivalent to hibernation. The Mohave ground squirrel, *Citellus mohavensis*, not only hibernates when the days are cold, but then reacts in similar fashion to heat and drought. Once again, as with hibernation, this small animal allows its temperature to drop, and therefore its metabolic needs to be reduced. The two forms of somnolence, the summer and winter kinds, can appear identical, but there are differences. Hibernating squirrels will not turn over if placed upon their backs whereas they will do so if merely aestivating.

Other examples of aestivators are gerbils, kangaroo mice and little pocket mice. Probably, with most of them, it is not so much the lack of food which prompts their aestivation, but the lack of water. To lower metabolism is to reduce respiration, and therefore to boost water conservation, with all breathing causing water loss. Another alternative is to behave more like a camel. This species can last for two weeks without water intake, and can lose up to 30% of its body water with no ill-effects. The dehydrated camel can then drink 30 gallons of water in ten minutes and return to normal. (A man who is given neither food nor water will die when he loses 15% of his body weight, and such a loss will normally occur within ten days, or yet more rapidly in hot conditions.)

In brief, animals and plants may not possess calendars, but are as aware of the seasons as are humans, if not more so. If seasonal happenings are amended because of human interference, the animals and plants will also have their expectations adjusted, as summers become hotter and drier, or wetter and windier, and the previous range of variability from one year to the next is nudged from the existing norm to something different. All life on Earth will have to adjust, or die, just as it has had to adjust over the millennia to every other form of climatic change.

That might seem acceptable, and therefore not disastrous, save that the human alterations may come quicker than the natural kind. Mankind's recent tinkering has not happened slowly, as it surely did when our paleolithic ancestors gradually learned about fire and conquest, about hunting and teamwork, about ways and means of greater survival. The past two hundred years have been explosive in their development. The growth in human numbers has been unparalleled, and greed – for absolutely everything – has been unprecedented.

The effects of this consumption have already been dramatic, with land clearance, with pollution, with increasing urbanisation and general alteration unimaginable a mere two centuries ago. The speed of this change was also unforeseen, bearing scant relation to the earlier routine of cyclical events, of warmer times and colder ones, of wetter years and drier, of happenings which were troublesome until normality returned. Life, in all its forms, has been in the habit of adapting, but in far more leisurely a manner than may soon be necessary. The drama to be unfolded as a result of human activity will certainly not be of concern solely to humans. Every other living thing will also be involved. Those millions of other species will find their lives in jeopardy, or merely altered, because one particular species has been excessive in promoting its aggressive, demanding and self-seeking form of livelihood in almost every corner of the world.

CLOCKS AND COLOUR

Periodicity – zeitgebers – cyclical growth – human 'trogging' – clock genes – pacemakers – hypothalamus – pineal eye – colour change – seasonal phases – diurnal, nocturnal – fiddlers – photoperiodism

'The periodicity . . . is to a certain extent inherited,' wrote Charles Darwin in 1880. Indeed it is. To a certain extent it is also imposed, as each of us is forced to acknowledge whenever we are abruptly transported to another continent. Upon our arrival elsewhere we are out of phase. There is still a longing to go to sleep, an enthusiasm to eat, and a period of greater alertness, but these are all happening at the wrong times – until our internal rhythms have had their clocks adjusted to the new location. This phenomenon of jet lag is a convincing demonstration of an age-old inheritance and, as further proof, it then occurs all over again when we fly home, this second translocation demanding similar readjustment before we can revert to our original routine.

There was frequent mention in the previous chapter of biological cycles, of triggers initiating some sequence of events, of sleeping and waking, of sexual regularity, of periodicity in its numerous forms. There are countless more. The dormouse not only arouses itself every three weeks or so from its hibernation, but always selects a time near sunset. Bees only arrive each day when certain plants have most nectar. Many biting insects only bite at definite hours. Fiddler crabs live in rhythm with the tides even when locked up far from the sea. Human babies go to sleep and wake up cyclically well before birth, this timing independent of their mother's own periods of sleep and wakefulness. Many flowers open their petals in readiness for the day, long before any brightening has appeared in the sky. Even some single-celled, and so-called simple creatures, like those responsible for the sea's phosphorescence, have a 24-hour rhythm to their luminosity. All such cycles are a response, built in or acquired, to the astronomical situation of days and nights, of seasons and of years.

They are a consequence of life on board a revolving planet which is orbiting its sun.

Biological rhythms were initially studied only by botanists. In 1729 a French astronomer-botanist reported that the leaves of a heliotrope plant opened and closed in a diurnal cycle. A century later Augustin de Candolle showed that these leaves still functioned cyclically even when kept in complete darkness. By the 1930s it was being realised that virtually every life form manifested some form of rhythm, and that three distinct components were involved. These were an input pathway, a pacemaker, and an output pathway to express that organism's reaction to outside events. In plant terms this meant a detector of the day/night routine, a clock (of some sort) regularly adjusted by the changing cycle, and a system for opening and closing leaves, for example, or for regulating photosynthesis.

In similar fashion the man-made timepieces we all employ, our watches and clocks, need some external input to set them going at the right time. They also need some means of adjusting their timekeeping and then, finally, they govern an owner's reaction to the information they provide. As we are not plants acting uniformly, but behave as individuals, our actual responses are up to each of us. Should we leave for work now or in 30 minutes, or should we eat, or do nothing of the sort?

With animals and plants this trio of systems can easily be identified experimentally. A group of starlings, caged outside and therefore subject to normal days and nights, had been accustomed to receiving food only at the cage's northern corner. The birds were then placed indoors beneath artificial lighting which, in its brightness and darkness, was arranged to be six hours out of phase with the outside world's day/night routine. After the starlings had become acclimatised to this new regime they were then returned to their former cage. Promptly they looked for food in the eastern corner, having had their clocks altered by 90 degrees, by a quarter of one day. This was proof that they did possess a timing mechanism, a means for its adjustment, and also a purpose in such a system.

Another experiment was planned to investigate the procedure whereby internal timepieces are adjusted. Humans, even if deprived of the objects strapped to almost every wrist or fastened conveniently to walls, can still be quite accurate about the time of day. So much is happening to give them clues. Quite apart from the sun's progress through the sky there are all the cyclical routines of life. Has the milkman called, or the postman, and has the bus arrived to take the neighbour's children off to school? Even without such promptings, and during sleep, some people can accurately time the end of their night's rest, without assistance from any clock serving as alarm. There is a marked increase of the hormone adrenocorticotropin one hour before waking, this acting as stimulant. The unconsciousness of sleep is therefore not as deficient in consciousness as might be presupposed.

Potato tubers were chosen for the investigation, it being assumed they would not encounter many clues, as humans do, about the time of day. The experiment would discover if there were cycles to the sprouting of potatoes, even if they were kept as free as possible from all extraneous influence. Therefore some tubers were placed within black chambers and were denied all light, noise or change in temperature. Such possible stimuli are known as zeitgebers, a name coined by Jurgen Aschoff, pioneer in biological rhythms whose earliest experiments were upon himself when he discovered a 24-hour variation in human heat loss. (Fundamental examples of his 'time givers' are the setting sun, decreased warmth, increased darkness and increased light.) With the potatoes installed in their monotonous conditions, missing every blatant indication of passing time, it was expected that all subsequent sprouting would occur quite evenly throughout each outside day.

Not so. The potatoes' rate of oxygen consumption, indirect indication of their rate of sprouting, was carefully measured. This showed that the new shoots were growing in a cyclical fashion, their growth always markedly different in the early morning between between five am and seven am. The experimenters, in hoping to control all outside effects, then realised they had forgotten about the slight but consistent irregularity in atmospheric pressure even earlier each day between two am and six am. Air pressure then goes slightly up and down, despite the much greater 'lows' and 'highs' when cyclones and anticyclones are taking their turn. It was therefore presumed the potatoes were detecting that regular daily difference and it was influencing their daily rate of growth. But, once again, not so. Even when atmospheric pressure was made absolutely consistent the tubers still sprouted slightly differently between five am and seven am. Therefore something else was at work, something not yet properly understood.

Wherever and whenever there is an efficient 24-hour rhythm it is axiomatic that a timekeeper must be involved. As a generalisation all internal clocks tend to slow down in the absence of suitable zeitgebers. It would seem that the clocks need prompting, much like winding, as well as actual signals, like checking against other clocks, to keep them running well. Their periodicity is 'to a certain extent inherited', as Darwin said, but needs – to a certain extent – to be adjustable. Perhaps a good analogy is with language. Humans are all born with a predisposition to speak, and all infants babble in similar fashion wheresoever they have arrived. This babbling ends after a few months, and they then begin to listen to the actual language of their region before learning how to speak it. Language acquisition, along with timekeepers, is therefore part inheritance and part environmental.

Human rhythms embrace far more than the cyclical events of sleeping and waking, of feeling hungry and so forth. Body temperature, always said to be 'normal' in healthy individuals as if unvaryingly constant, ranges over 1.5 degrees C. in any day, being lowest before dawn and highest in each day's latter half. The excretion of urine is also rhythmic; so too pulse rate, blood pressure, oxygen uptake, intestinal activity, and therefore, it is assumed, every other form of bodily function.

Even the fundamental chemistry of metabolism fluctuates evenly each day. Excretion of sodium and potassium rises and falls according to an established pattern, with maxima and minima independent of working hours, of resting periods, of meal times. Instead both of them peak at about midday and reach a minimum at midnight. If people are transplanted from their homeland to another continent their excretion rate remains on home-time for a while, but is slowly adjusted to the local midday and midnight, with sodium the first to be altered and potassium a few days later. So, to use a modern phrase with more precision than is customary, what makes us tick?

Unlike the actual clock inside Captain Hook's crocodile there is no mechanical timepiece within the human frame, but the hypothalamus comes very close – and more about that later. The ticking, the pulsing, the cyclical activity is a feature not only of the organism as a whole but of its component parts. Heart tissue, for example, beats/contracts rhythmically. Its haste is adjustable, as we all know with our pulses speeding up when need be, but the heart's beating is a fundamental property of its individual muscle fibres. As a group they continue to contract and relax for the 2.5 billion times of an average lifetime before, perhaps starved of oxygen during an attack, they are compulsorily halted.

However inexplicable, or even incomprehensible, is this steadfast pumping it should not be too surprising that other whole organs or groups of muscles also have such periodicity. After all even single-celled animals do so. *Euglena* and *Paramecium*, for example, the one-celled stand-bys of much school biology, have been shown to possess endogenous rhythms. So too primitive multicellular organisms, like algae and fungi. And so too, which is possibly yet more surprising but ought not to be, single intestinal cells taken from the human intestine. If maintained on their own by ordinary tissue culture methods they still follow the same rhythm as their compatriot cells which form part of a living *in situ* human intestine. It is rather like trying to find the sparkle of a diamond. The more a diamond is fragmented the more pieces are thereby created which also sparkle.

Humans have experimented upon their own daily cycles to discover more about these basic rhythms. As if wishing to emulate the potatoes encased within dark chambers, certain experimenters have installed themselves inside deep caves. David Lafferty, for example, spent four months within a Cheddar cavern in 1966, before thinking it was only 7 July when a friend came to fetch him, as prearranged, on 1 August. His 'days', from one awakening to the next, had lasted from 19 hours to 55, his longest 'night' being 37 hours. In time within his self-imposed in-carceration he became more normal, his 'days' lasting from 24 to 27 hours. He spent, understandably in that silent and surely boring predicament, 60% of his time asleep, as against the standard human pattern of 33%. A greater achiever in this sport of 'trogging' was Milutin Veljkovic. He lived underground from 24 June 1969 to 30 September 1970, a considerable proportion – 1.8% – of a normal lifetime. Michel Siffre, yet another human mole, spent a mere 62 days below the Alps in 1962. He was much more regular than Lafferty, but awoke later and later each day. He was due to emerge on 14 September, but eventually had to be retrieved because, in his isolation, he still thought it was only 20 August.

Waking later and later each day seems to resemble those of us who are referred to as owls rather than larks. Early birds are the individuals who, often irritatingly to others, like to start the day at cock-crow. Most likely to be annoyed are all the owls who prefer to stay up late, and then to get up late, and certainly later than the dawn chorus of irritating larks. It would seem that the early risers have a genetically endowed and endogenous rhythm nearer 23 hours while the night-owls would prefer a 25-hour day.

Indeed a 'clock gene' has been discovered. This differs between individuals, favouring either cytosine or thymine as part of its fundamental make-up. At Stanford University the relevant genes of 410 individuals were examined, and 191 were found to be of the cytosine variety. These people, in general, favoured later hours for work and exercise, while the 219 others performed better in the morning. Both, of course, are compelled to adjust their personal, and possibly gene-influenced, wishes to the earthly situation, namely that days are 24 hours long. It is not possible for individuals to operate either a 23- or 25-hour cycle when the rest of the world, and the planet itself, is working on a 24-hour basis.

Despite the vagaries of long-term cavers, and the human division of larks and owls, it would seem from recent work at Harvard Medical School reported in *Science* that our bodies are actually very good at keeping accurate pace with the Earth's rotation. The average day for 24 individuals who lived for a month without clocks or any sighting of the sun proved to be 24 hours and 11 minutes. There was no difference between the old and young, as had formerly been believed. The volunteers were deliberately upset from their normal rhythms by having to go to bed either four hours earlier or four hours later each night. Even so their bodies were able to 'know' the time, save for that modest average error of 11 minutes in every day.

These natural clocks can be stopped experimentally. Two sets of bees were once given different treatment, one group permitted to live normally and the other chilled to four degrees C. for six hours. The control bees continued to be most active at their usual time, from 10 am to 1 pm. Those which had been chilled did not start their search for food until the afternoon, reaching a peak of activity at 3 pm and carrying on until 4.30 pm. The six hours of cooling had therefore postponed the peak by 4.5 hours, and had also made the feeding period rather more erratic. The bee clocks had therefore been stopped, as well as deranged. Such an experiment makes sense. It seems reasonable that chilling halts the system, whatever it is, but certain other experiments can seem far less reasonable, both in their actual purpose and their subsequent discoveries.

As reported in *Science* in 1998, 15 people were made to experience a bright light shone on the backs of their knees during their night's sleep. Astonishingly this treatment adjusted their biological clocks. The presumption that light detection was a matter solely for eyes therefore needed further investigation. It was already known that many animals can be influenced by light, even if without eyes. Fruit flies, the stand-by of many a laboratory, can do so whether or not they have vision. Even amphibians, reptiles and birds have been shown to possess sensors deep in their brains which react to light penetrating the skull. Mammals are different. Without eyes, or so it had been thought, light had no effect. There is also the conundrum, which is possibly irrelevant, that nocturnal animals sleep most when the light is brightest.

It was subsequently learned that the clocks of mice could be adjusted even if the seeing part of their eyes was entirely absent. Eventually, after much scepticism from various workers in this field, it was discovered that certain molecules known as cryptochromes were the light detectors. These have been shown to exist in a wide variety of species, from a weed called thale cress to fruit flies, and also to mammals. During this research fruit flies were found to have light-detecting clock mechanisms all over their bodies, from wing to thorax to abdomen and also internally on their intestines. As for mammals, even their skin cells growing in a culture dish have been shown to possess pacemakers. All sorts of tissues – liver, muscle, testes – have also been found to contain endogenous rhythms. A Harvard neurobiologist recently summed up current thinking: 'If you haven't found rhythms in some tissue you haven't looked hard enough.' The extra-ordinary knee experiment therefore not only makes some sense but its results could be invaluable. Many a night-worker, a shift-worker or global traveller would be extremely interested in an effective system for adjusting their clocks, something more user-friendly than being woken up to glare at light bulbs which is the current way. A light behind the knee could, apparently, do the trick just as well, and much more pleasingly.

Apart from the numerous timepieces within mammals there is a central clock, a Big Ben among the lesser forms of ticking. This exists, as already indicated, in the hypothalamus. 'Is there any pie . . . into which the hypothalamus does not dip its finger?' asked a leader in the *Lancet*. In humans it occupies only one three-hundredths of the total brain mass, but does perform an inordinate quantity of roles. One of these is now known to be the master clock, and its centre lies in an area of the anterior hypothalamus known as the SCN, the suprachiasmatic nucleus. It consists of about 10,000 very small neurons, some of the smallest in the brain, and they secrete hormones, such as vasopressin. More to the point, in this chapter on timekeeping, they do it rhythmically. They also send rhythmic signals to the pineal gland. The SCN importance was emphasised when such cells were taken from a rat's brain, kept in culture, and then transplanted into other rats with faulty SCNs. Within two days the recipient rats had their natural rhythms restored.

At once the system becomes both more comprehensible and more complex. Birds, reptiles and fish also possess a pineal gland, this being a minute mass of nerve tissue deep between the cerebral hemispheres. In former ages, and with simpler organisms, it served as a primitive central eye, responsive to light. Even in more advanced creatures it still contains light-sensitive cells, with most of the gland's current function poorly understood. René Descartes called it the seat of the soul and, if a time-keeper is considered to be the controlling centrepiece of any system, he was not far wrong. The gland secretes melatonin, but only in darkness, this hormone being directly involved, along with serotonin, which is widely distributed around the body, in regulating the day/night sleep routine.

In short, and cutting corners, the pineal gland was an early light-detector, and has still not yielded its ancient role. Solar radiation, and the steady alteration from daytime to night-time, still lies at the base of all activity. The hypothalamus, so widely involved in almost everything, still influences the pineal gland in its age-old activity. Vasopressin from the hypothalamus constricts blood vessels, adjusts kidney output, controls water in general, and does so in a rhythmic fashion. Melatonin, coupled with serotonin, looks after our basic routine. And finally, with so much rhythmic influence in every sphere, each cell in the body – or seemingly so – pulses in sympathy.

As analogy the global community of humans behaves in similar style. It would be quite possible for some casual observer not to notice any clocks or watches, but it would be quite impossible not to notice the rhythm of our ways, the getting up and going to bed, the going to work and coming back, the daily labouring and then the resting hours. There is a master rhythm, controlled by Earth's spinning on its passage round the sun, and there are all the lesser rhythms from each and every one of us, the single cells in this worldwide web of complex interaction.

Days and nights have existed ever since life began. It is therefore small wonder that this fundamental cycle has embedded itself deep within every living thing. The clocks are everywhere. And so are the zeitgebers, the givers of time to keep them accurate. Everything responds to everything else, or so it would appear, and seems most reasonable. Trees, of course, are affected by light, by temperature and by humidity. They also, according to some very recent work, respond to tides. The diameter of their stems can be correlated, if all other influences are denied, with tidal ebb and flow.

It is not surprising that the moon, although a lesser object than the sun, also plays its part. It too has been in existence, as nearest neighbour, ever since life began. The fact of one planet with a single satellite, these two circling each other and both orbiting the sun, lies at the root of everything. Every rhythm and cycle, every ticking and oscillation, every form of periodicity, whether inherited or acquired, is subservient to the astronomical situation. That is the basis of our time.

Colour change Victorian naturalists who were also painters, as they were so often, took great delight in cryptic colouration. Peacocks were depicted with their blue heads against the azure sky, their multi-tinted bodies against appropriate vegetation, and their amazing tails against no less astonishing backgrounds. The birds themselves were therefore virtually invisible, save by discerning eyes. The painter-naturalists were equally delighted by Arctic foxes, mountain hares, stoats and ptarmigan, splendidly white on wintry snow and then more richly coloured among summertime's exuberance. The painters tended to disregard the polar bear, the snowy owl, Greenland falcon and American polar hare, all equally white in wintertime and no less white in summer.

Other northern creatures were also omitted in these landscapes, such as the moose, musk-ox, glutton, reindeer and raven. Their colouration blends reasonably with surrounding tundra in summertime but conspicuously stays that way after snow has fallen everywhere. Not one of these animals turns white. Yet more unhelpfully, for any generalisation about colour adaptation, the Hudson Bay lemming alters its colour annually but the Norwegian lemming fails to do so. As for the Arctic fox this single species has two phases. One does indeed alter from brown in summer to white in winter, while the other changes from grey in summer to black in winter.

Plainly, as both phases exist, with the non-changers surviving apparently just as well, there is more to colour than simple adaptation to the seasons. Despite the terrestrial landscape changing from browny-grey in summertime to creamy white in winter, this being the most dramatic seasonal change on Earth, there is no universal law that the local animals must follow suit. A snowy owl when standing on tundra flatness can hardly be more conspicuous during the warmer months of every year, being possibly the only item of note for miles in each direction. Similarly, although a black Arctic fox is less blatant in summertime, it is black on white in winter. As for the animals which do change when summer warmth becomes winter cold their timing is generally imperfect, with white Arctic foxes running around most visibly long before any snow has fallen.

In other portions of the globe there are other seasonal changes. Deciduous forests become bare poles in wintertime, dry land becomes wet, and even the drab grey of deserts can burst into flower, but colour change by animals in response to these other alterations is extremely limited. Fallow deer, *Dama dama*, have dappled spots in summer, much like the young of many deer, but they lose them in winter. Japanese deer, *Sika nippon*, are similar, with every adult and all youngsters losing this attractive dappling when, or even well before, the leaves depart. Countless insects possess adaptive colouration, blending beautifully with their favoured backgrounds, but insects tend not to be around during wintertime. They pass that season as eggs, or pupae, or well concealed within convenient hiding spots. Tropical insects are often duller in their colouration during the drier time of year. As for all the grasshoppers, butterflies and even moths whose bright underwings give a flash of colour when they depart, these creatures usually lose that brightness at the rainy season's end. One suggestion for this change is that insects are then rarer, their predators are hungrier, and cryptic behaviour then pays greater dividends than the startle effect of all those underwings. The same applies to aposematic insects in the temperate world, those with warning colours such as wasps and ladybirds. The surviving adults hide in wintertime, the advantages of conspicuity having no longer such merit.

The fact of summer becoming winter in all its forms is one astronomical circumstance. The second such basic fact is that daylight ceases, being replaced by night. Nocturnal animals must either be hidden during daytime or be experts at camouflage. Nightjars and woodcocks are the supreme concealers among birds, best within vegetation but even on open ground. Bizarrely the desert varieties cannot be seen until they are almost underfoot. Whereas snowy owls stand out like white posts in tundra summertime, most owls have a colour in keeping with their locality, being sandier in the deserts and tawnier among trees. Barn owls are particularly visible and therefore retreat in daytime, but the screech owl of North American forests is one of those which do not hide. Instead, and cunningly, it not only makes slits of its otherwise conspicuous eyes, but shrinks its form by drawing in its feathers.

Most birds are diurnal rather than nocturnal, and a few of the night variety even spend their days in burrows, such as New Zealand's famous kiwi, but all the birds which have to incubate their eggs without benefit of holes or within the security of large communities must behave as if they are nocturnal and attempt to hide. As a generality the more colourful partner, usually the male, is the one *not* doing the incubating. Emphatically making this point is the painted snipe, where the males are drab and the females gaudy, it being the males who carry out the incubation. Eggs are vulnerable as well as the parents sitting on them. In general, although there are exceptions, eggs tend to be white if they are laid and incubated out of sight of predatory eyes. Those laid in the open are more colourful.

A wonderful creature is the fiddler crab, such as *Uca pugilator* and *Uca minex*. Each species incorporates the two most dominant astronomical factors into its lifestyle, the two which humans have tried to blend, so unsuccessfully, into their calendars, namely the phases of the moon and the daily disappearance of the sun. These crabs spend much time in their sandy burrows and emerge not only when the sea has retreated but when the surrounding sand has sufficiently stiffened to prevent burrow collapse on their emergence.

This correlation of behaviour with the tide-controlling moon is also coupled with the sun's presence or absence because each crab's colour spots, its chromatophores, vary in size according to the time of day when the animals leave their burrows. This solar rhythm is engrained because the spots do not alter, like a chameleon's colouration, in immediate response to the prevailing conditions. The expansion and contraction of crab chromatophores in tune with the sun's presence or absence will persist for some 30 days even if the crabs are kept isolated from external influence. Lunar and solar cycles are quite unrelated, with night becoming day having no relationship with the moon's effect upon the tides, but the crabs take all that in their stride. They emerge suitably coloured to meet the day or night when the tidal flow is at a convenient stage.

A brief word about human colouration. Dark skin is commonest near the equator, brown skin is most frequent in the warm temperate latitudes, and light skin further north. Despite this broad truth no one has suggested that human colour variation is a form of concealment, with the whiteness of snow more probable nearer the poles and darkness a sombre feature of tropical forests. Neither does human colour make much sense with regard to warmth and cold. Dark skin absorbs more heat from sunlight than does pale skin, and a dark-skinned individual loses more heat during a cool night than does a pale-skinned person. It is possible that blackness acts as a shield against the harmful side of solar radiation, but most sunlight occurs well to the north or south of the equator – where browner people live – where the sun is high in the sky during summer and the daylight hours are much longer. It is also a fact that dark skin prevents excessive manufacture of vitamin D, and that may be of greater year-round performance. Or there may be some other cause, or group of causes, quite independent of all such reasoning.

In short, colour often involves a host of possibilities, only one of which, and a minor one at that, involves adaptation to the changing seasons. Concealment is another factor, however many exceptions there are to this further attribute. Finally, despite the elegant brushwork of those earlier naturalists blending magnificent birds within their luxurious habitat, it is easy to believe that any predator of peacocks experiences little difficulty in locating its colourful prey. As for peahens, the much less exuberant females of this species, they never blended with their backgrounds half so well, and were usually omitted from the paintings.

*

Plants It may seem odd encountering plants after animals, with plants so straightforwardly dependent upon the sun, but much more work has been done on animals. Besides, plants cannot migrate, become nocturnal, hibernate or burrow underground. Nevertheless there are some intriguing adaptations to changing climate. Whereas cold-temperate trees, in general, shed their leaves in winter and grow in summer, the hot-temperate world is contrary. Photosynthesis around the Mediterranean, for instance, is a wintry business, along with growth. Summer is a time for dormancy. Similarly, although leaf-shedding is a more northerly matter, the most northern trees of all do not shed their leaves. Conifers, save for the deciduous larch, maintain their needles.

It had always been assumed, perhaps since hunter-gatherers watched how plants leaped ahead in springtime, flourished in summer, and produced their fruits when autumn came, that warmth served as trigger for all this activity. Not until 1920, despite the absolute reliance of botany on sunshine, was it realised that day length, now called photoperiodism, sets the process going. The length of the light period is more important than the intensity of light, with day length being the only accurate indicator of the progress of each year.

It was W.W. Garner and H.A. Allard, both of the US Department of Agriculture, who learned of day length's crucial role and who, after working with numerous species, realised there were three types of plants. These they named as short-day, long-day, and day-length neutral. The short-day variety are stopped from flowering by long days and short nights, and the long-day kinds are converse, never flowering when there are short days and long nights. As an example, the cocklebur, *Xanthium pennsylvanicum*, will steadfastly fail to flower if given 16 hours of light and 8 hours of darkness. However, if this plant is given but a single darkness dose of 9 hours, or more, followed by 15 hours, or less, of daylight it will immediately leap into floral activity. More amazingly it will continue to flower for a year however much daylight it is given following that single dose. A long-day variety, by contrast, is *Hyoscyamus*, the tobacco plant. It needs 10–11 hours of daylight, and will continue to flower so long as daylight is as long, or longer, than that quantity. The general word trigger is therefore entirely accurate, it initiating the process whatever this may be.

It might have been expected, indeed it used to be expected, that the apex of a plant needed to receive the stimulus of light. After all, it is the apex which does the growing. Later it was realised that leaves, and not the growing tip, need the stimulus. Perhaps everyone should have remembered animals. Light does not have to shine upon their gonads to get them going, but upon their eyes. Later still, and following experimentation on different plants, it was learned that only one leaf, or even a portion of one leaf, needs to be stimulated for the whole business of efflorescence to occur. Finally, it was discovered that the leaf could be detached, then stimulated, and then grafted on to another plant for that secondary plant to begin activity. This grafting need not even be between a single species. *Hyoscyamus niger* is a long-day tobacco plant and if a single, but stimulated, leaf from it is grafted on to the Maryland Mammoth, a short-day tobacco, this can be induced to flower in long days. And so too vice versa.

Inevitably, therefore, a hormone is involved. Oddly, unlike other substances, such as sugars, which also travel through a plant, hormones proceed more slowly. Also, although photosynthesis operates more effectively if light is strong, the quantity of light needed to act as trigger can be extremely modest, such as one foot-candle. (This quantity is what it says, namely the amount of light from a standard candle at a distance of one foot.) Compared with the full vigour of sunlight a foot-candle of illumination is negligible, but some plants can be triggered to adjust their flowering cycle even by 0.1 of a foot-candle, which is only twice as bright as bright moonlight. Human beings, with eyes so adept at permitting vision both in bright sunshine and sombre moonlight, should remember that sunlight reflected from the moon is pathetically weak compared with direct solar power. We cannot even see colour by moonlight, and should be astonished that its strength is sufficient to trigger any kind of plant activity.

Of course warmth is important for plant growth, and so too moisture, but day length is the one reliability concerning the progress of each year. Animals know that, and plants know that – however wrong it may be to attribute them with knowledge. Early humans certainly knew that, with so many of their megalithic monuments attuned to solstice and equinox, the critical marker points of Earth's progress round the sun. And modern humans, obeisant to their calendars, run lives in accordance with the annual succession of days, of weeks and months rather than the actual weather, which may be wetter, drier, colder or hotter than is expected for that season of the year. We too are organised by day length even though, with artificial lighting so pre-eminent, we are kept aloof from the changing length of day.

To every living thing there is indeed a season. A world without seasons and without nights and days is unthinkable. To the animals and the plants these cycles are paramount. To all early people they were equally crucial. To some alive today they have lost none of their authority. Even urban individuals, so artificially maintained, are still ruled by them, sometimes subtly, often blatantly, however much they may resent, deny or welcome the fact. As planes hustle us across the lines of longitude, or lights turn night into day, we could do well to think rather more about the basic cycles of our planet. It is strange that humanity pays so little attention to its own internal rhythms. To everything there is most positively a season. There are no exceptions, not even for impetuous, arrogant, technological and modern *Homo sapiens*.

Chapter 11

NATURAL FLUCTUATION

*El Niño – 1997-8 – Knock-ons – ENSO – Forecasting – La Niña – MJO and PDO
– warm pool – North Atlantic Oscillation – statistical partiality – 1789*

What is now known, only too well, is that the El Niño event of the South
Pacific Ocean disturbs climate right around the world. What is also known
is that La Niña, so frequently its successor, disturbs climate but in contradictory fashion. As third point it is known that the Southern Oscillation, an
atmospheric see-saw over the equatorial Pacific, is linked to El Niño, the
whole phenonemon being known as ENSO, the El Niño Southern
Oscillation.

This distant disturbance is not some irrelevant event, of interest solely to
global oceanographers and South American fishermen, but has been
defined as 'the strongest natural interannual climatic fluctuation'. It is
therefore of considerable consequence, but what is not known is yet more
basic than the fact of its occurrence. Where and what is the trigger for
ENSO, and why are particular years blessed or cursed with the event? A
further query of similar, or perhaps greater, importance is whether ENSO,
plus all its dependent changes, is being influenced to any degree by
mankind's activities.

The event itself results from the nutrient-rich cold water of the coastal
Humboldt current, which surges northwards off western South America,
abruptly being replaced by eastward-flowing and nutrient-poor warm
water from the equatorial Pacific. Alexander von Humboldt hardly
discovered this current, it being well known to every local fisherman, but
he did set sail upon it in 1803 when travelling from Guayaquil to Mexico
and he did, as was his way about everything, take measurements of its
properties, including its temperature. In the fickle manner of history this
single piece of work, a modest portion of Humboldt's five-year stint
exploring South America, has become his most famous monument. When
the current's warmth switches from cold to warm during an El Niño phase

this can mean a rise of four degrees C. or even higher in its surface temperature.

The increased warmth encourages low-pressure systems and rising air, and the warmer air then travels westwards to descend by south-east Asia. At a lower level it then returns towards Peru, a complete reversal of the normal situation. Within the sea the warmer water flows much further south than usual, superimposing itself above the colder water, thereby killing off much of the marine life. This switch means less scouring of the sea-bottom to bring up nutrients, and therefore much less plankton food for fish. The fish suffer from this lack and sea-birds therefore suffer from a lack of fish, with human fishermen suffering in similar style.

The anomaly has long been known to many South Americans even well away from the fishing ports, its effects often disastrous for national prosperity. The bad years have been officially listed since 1877 when weather in that region was first properly recorded. They have been scientifically examined since 1892 when formal academic interest in the phenomenon was initiated, but not until the strong 1972–3 event had taken place was such endeavour spurred into greater action, and not until then was outside curiosity attracted to this Humboldt happening. The associated collapse of the Peruvian anchovy fishery, which had been booming since the mid-1950s, focused political minds as well as those of scientists and fishermen.

No sooner had this work been accelerated than ENSO, as if in response to the spotlight of enquiry, not only became a more frequent occurrence than its traditional arrival every three to seven years but grew much more severe. The El Niño of 1997–8, called by some 'the climate event of the century', is estimated to have caused $33 billion in damage and to have killed 23,000 people worldwide. 'Small earthquake in Chile; not many dead,' the journalistic in-joke about the unimportance of foreign occurrences, has already been mentioned. 'Small temperature change in Pacific waters; many thousands killed worldwide' now proclaims a greater and more disturbing truth.

The extent of the 1997–8 oscillation was quite unprecedented in every way. According to the Pacific Marine Environmental Laboratory in Seattle it 'developed so rapidly that every month between June and December 1997 set a new monthly record high for sea-surface temperatures in the eastern equatorial Pacific'. Before that momentous year, according to the same report, 'the previous record-setting El Niño occurred in 1982–83 ... These two "super El Niños" were separated by only 15 years, compared with a typical 30–40 year gap between such events earlier this century.'

Record ocean temperatures are naturally important, but the knock-on effects elsewhere were overwhelming. Peru itself lost 30 important bridges from torrential rain during that momentous season of 1997–8, together with 370 miles of major road. In Chile 80,000 people lost their homes, and even some of that country's Atacama desert received rain, an event not previously recorded in that region since the Spanish arrival. Much further north the city of Acapulco received 16 inches of rain in 24 hours. Hurricane Pauline killed 400. Texas suffered extreme heat, with 85 deaths as a result. Laguna Beach, California, experienced weather it had never previously encountered. There was snow in Mexico, and an 'ice-storm' in Canada which blocked off half of Montreal's electricity for a week. If such misfortunes were indeed the result of a band of Pacific ocean being warmer than usual the change was not only remarkable and scientifically intriguing but had much to answer for.

The ENSO of 1997–8 was also blamed for disastrous fires in Indonesia, the worst on record, and for a similar spate of conflagrations in Amazonia. Both furnaces were, to a large extent, people-induced, but the tinder-dry situation enabled the fires to be exceptionally horrendous, with polluted air making breathing and visibility difficult hundreds of miles away. Even Kuala Lumpur and Sarawak, both in Malaysia, reported a 'dramatic increase in respiratory disease' resulting from those Indonesian fires. The Sudan and the Philippines suffered debilitating famine from lack of rain. Contrarily much of eastern Africa was drenched.

Just as one man's meat is said to be another's poison so is one area's flooding another's drought. There cannot be universal wetness any more than there can be a worldwide lack. Moisture collected from the oceans has to fall – somewhere. What is so disruptive in an ENSO year is the alteration to traditional patterns, and to general expectation. Heavy rain in a desert can be as damaging, via floods and sudden waterways, as its lack in agricultural areas. East Anglia, Britain's driest region, is a green and pleasant place, but would be appalled, and drowned, if it received typical Indian rainfall. India, so often beset by drought, would become largely uninhabitable if it only received East Anglia's precipitation. An absence of rain anywhere will lead to failed crops and possible starvation. An unexpected deluge can be equally devastating, by carrying away bridges, destroying roads, demolishing homes and killing people.

Broad rainfall patterns do occur in association with an ENSO year. Land areas likely to be excessively wet are:

Zambia, Zimbabwe, Tanzania, Madagascar (from November to May);

India, Bangladesh (June to September);

most of Australia, Indonesia, New Guinea, Philippines (November to May);

northern South America (July to March);

Peru, Bolivia, and southern Brazil (May to April).

Conversely the excessively dry areas are likely to be:

southern Sudan, Uganda (October to April);

southern India, Sri Lanka (October to December);

Texas, Louisiana, Alabama, Florida (October to March);

Wyoming, Colorado (April to October);

northern Chile (June to November);

Uruguay, Paraguay, and northern Argentina (November to February).

Of course rainfall cannot be obedient to national or state frontiers, and those generalities are not as precise as the local citizenry might wish them to be when estimating its chances for drought or deluge. However true this broad pattern, and however similar are the knock-on effects of ENSO behaviour in earlier years, there is no guarantee that future behaviour will be in similar style. As everyone on Earth knows, everyone who is over the age of, say, three, the presence or absence of rain is a most fickle business. Similarly, and following a Pacific Ocean warming, the consequent likelihood of excessive wetness or excessive dryness in any particular place is only broadly equal from one El Niño to the next.

The single certainty is that an ENSO season will more probably be aberrant, all over the globe, in all manner of ways, and often disastrously. Over the years human beings have been extraordinarily adept in their styles of livelihood, making a living in deserts, in less arid areas, in forests, on ice, and in regions with hundreds of inches of rain each year. What they find difficult, or downright impossible, is suddenly having to cope with a totally unexpected and absolutely abnormal kind of weather.

As extra cause for concern the ENSO phenomenon is on the increase. There were three extreme events during the 30–year period from 1950–80, but there have been four major El Niños since 1984, including the longest on record which lasted from 1990 to 1995. The actual title El Niño is something of a misnomer. It arose because South American,

Spanish-speaking fishermen traditionally noticed the event at Christmas-time. They therefore named it for the Christ-child, the young boy, but its consequent deluges, droughts and so forth take place during every other season of the year, often far removed from Christmas. As for La Niña, with its effects so contrary, its name of young girl formed an easy counterpoint to young boy.

The El Niño phenomenon was influential around the Andes, and far from the coastline, long before it acquired its famous name. No one knew it at the time but there was a link between the ocean's changing temperature and the right season to plant potatoes. Farmers in drought-prone Andean regions, such as the Incas in earlier centuries, traditionally observed the apparent brightness of stars in the Pleiades to decide when their most important crop should be planted. The less bright years indicated a probable drought during the growing season. Therefore plant later and hope to get a better crop. Today's scientists, notably from Palisades, New York, reckon that the poor visibility is caused by 'subvisual high cirrus clouds' and these, in their turn, are linked to El Niño years. Plainly the temperature switch of the Humboldt current is not a new occurrence. What is novel is a better understanding of the far-reaching effects of this oceanic alteration.

ENSO's abilities to cause catastrophe and make climatic headlines have become famous, but there are more insidious effects which are less immediately apparent. Even slight changes in rain or temperature can crucially affect the survival of pathogens, of vectors, and the reservoir species of infectious disease. According to the World Health Organisation many areas experience a 'dramatic increase' in malaria during El Niño times, with 'quantitative leaps' recorded in Bolivia, Colombia, Ecuador, Peru, Venezuela, Rwanda and Pakistan. In Sri Lanka the incidence of malaria 'leaps four-fold' in any El Niño year. Rift Valley Fever, which principally affects livestock, but also its dependent people, was exacerbated by rainfall which was '60 to 100 times heavier than normal' in southern Somalia and northern Kenya. 'Virtually uninterrupted' rain falling between September 1997 and January 1998 was thought to have been responsible for the fever epidemic.

Rain is not always to blame for high malaria years; low rainfall is more important in Venezuela, and probably Colombia and Guyana. As mosquito larvae need water in which to breed, the association between dryness and this disease is perplexing. Speculation has focused on a reduction of the mosquitoes' natural predators during dry times, such as other insects and amphibians, and on a lessening of the human population's immunity in periods with smaller quantities of rain. No one yet knows the answer, however well known it is that El Niño times are partnered by more malaria.

There is also an El Niño association between cholera and excessive rain. Tanzania suffered 40,249 cases, with 2,231 deaths, in the wet year of 1997, as against 1,464 cases with 35 deaths in the dry time of 1996. Cholera was also rife in Peru, Bolivia, Honduras and Nicaragua during 1997 as a result, it is believed, of El Niño's effect upon the weather. During the first three months of 1998 Peru alone experienced 16,705 cholera cases and 146 deaths. Comparable figures for that trimester were 11,335 and 525 in Uganda, and 10,108 and 507 in Kenya.

What WHO, and others, would like to see is a forecasting system so that preparations can be made. Since 1985 the Tropical Ocean Global Atmosphere programme has been in place. It promptly predicted an El Niño for 1986, which did indeed occur, but there had not been universal faith or optimism in the forecast. According to a team at Columbia University, New York, there were only a few successes. Prediction of drought in north-eastern Brazil caused its state officials to anticipate hardship and thereby it planned to mitigate the damaging effects. The Ethiopian government also took appropriate steps when drought was forecast, but there were failures to counterbalance such successes. Warned of a severe drought in Zimbabwe the farmers planted fewer crops. Unfortunately, or fortunately, the drought was not so severe as had been predicted, but the unwillingness to plant crops meant that food production was severely less than it might have been. In Australia the predicted drought did arrive, but the modest rain which fell, and had been anticipated for its modesty, happened to arrive at the very best time for crops. Therefore production was not so bad as had been feared. Peru received good early warning of the 1997–8 El Niño but severely underestimated the event's magnitude and it therefore suffered in consequence. Such misjudgement does seem entirely understandable, with that particular disruption being 'the climatic event of the century'.

As for La Niña this 'girl-child' often, but not always, arrives straight after its opposite. The alternate event refers to the extensive cooling of the central and eastern Pacific, and to a general lowering of temperature, notably in the tropics. El Niño and La Niña are therefore the two extremes of the Southern Oscillation. Those parts of Australia more likely to suffer drought during an El Niño time are usually wetter than normal during La Niña. Despite, or because of, the southern and tropical cooling there are likely to be temperature rises in China and the US. This increased warmth may be the reason why there is believed, by some, to be more hurricane activity in the Caribbean and the US during La Niña times. Nothing can happen climatically, or so it would seem, without it affecting something else, either boosting it or downgrading it, or causing some other change. Certainly nothing can happen during the 'strongest natural climatic fluctuation' without massive change elsewhere.

Hence a wish to know what triggers each El Niño, assuming that a single trigger is at work. One curious 'source of weather', as it has been called and may be relevant, is the MJO – the Madden-Julian Oscillation. This atmospheric disturbance, which originates over the Indian Ocean, vacillates every 30–60 days and may give a nudge to ENSO just at the right time. The last straw does break a camel's back. All earlier straws merely create the state when one more straw, one more nudge, will cause the previous stability to collapse. MJO, it so happens, was particularly active in late 1996. Trade winds in the western equatorial Pacific had been stronger for a year, and ENSO warming of the eastern Pacific was therefore more likely, with the MJO behaviour, possibly, acting as the final straw.

There is also PDO, the Pacific Decadal Oscillation. This further natural vacillation of ocean and atmosphere in the region has a periodicity, not of a month or two but of several decades. In tune with PDO, and its lengthy fluctuations, it so happens that the tropical Pacific's temperature has been slightly higher than normal since the mid-1970s. And it is since the 1970s that there have been not only more El Niños than normal but two of the 'super' El Niños as well. PDO may therefore be relevant, whether or not it is a basic cause.

A third possibility is whether some of the catastrophic events, such as earthquakes and volcanoes, themselves trigger or influence the ENSO pattern. Dust and ash spewed into the atmosphere certainly affect climate, but there does not seem to be correlation between the major eruptions and El Niño years. Another, and entirely converse, option is whether the sea level changes linked to ENSO years have helped to cause eruptions. Some correlations have been found, but there is plenty of dispute. In any case, with all the inter-relationships which exist, there may not be a single ingredient which directly causes the ENSO phenomenon. PDO, MJO, volcanic dust, early or late monsoons, and every other happening may all play their part, even if the part they actually play is bewilderingly obscure. No one knows which camel straw will be the last, the back-breaker, the most significant.

Human influence, and global warming, are a further possible cause, not of ENSOs in general, which have an ancient history, but of the recent spate of exceptional events. Similarly, and once again contrarily, the ENSOs themselves may have added to the worldwide warming. Global temperature usually rises by a small portion of a degree when the warmer Pacific of an ENSO adds some of its heat to the atmosphere. Therefore, putting two and two together, the increasing warmth of the past century may have helped to cause more ENSOs, and more ENSOs may have helped to increase the global warmth.

Computer models are being increasingly used to discover past relationships, whether a change in an ocean current did lead to warming, or whether some other warming led to change in an ocean current, and on and on. Only when there is better understanding between causes and effects will predictions become more accurate and more influential. The boom anchovy years of Peru, which collapsed so dramatically in keeping with the 1972–3 El Niño, have now been reinstated. During 1997 that country's commercial fishing generated 4% of the gross national product and $1 billion in foreign exchange. Such a major industry, so closely linked to a changing nearby current, would greatly welcome forecasting. So too the more distant economies when bad years or good years can break or make millions of livelihoods. Should more corn be planted in Zimbabwe? Should deforestation be discouraged in Indonesia? Should famine relief programmes be instituted ahead of time to prevent awesome loss of life? And, if the forecasters prove to be wrong, who pays, save for those who die as a consequence?

If El Niño was more reliable everyone could be happier. If its effects were consistent that too would also be a blessing, with suitable preparations being able to divert some of the agony. Unfortunately not only are the occurrences irregular but so are the results. For example, the so-called 'warm pool' of ocean water lying to the north of Papua New Guinea was recently shifted hundreds of miles to the east as a result, it is thought, of El Niño. Something of the sort happens in every El Niño year, but in 1997 the effects were different. Papua, usually deluged with rain, suddenly experienced a six-month drought, the worst for half a century, resulting in poor crops and a damaged economy. For some reason the warm pool had moved even further east than usual, with the March cyclone named Justin possibly to blame. Or maybe the fault was El Niño's. Or maybe it was the low-pressure system which traditionally builds over Tahiti. Or maybe something else, or several different kinds of something else.

The publicity accorded to El Niño has caused it to be the scapegoat for every climatic anomaly (having taken the place of witches in some earlier centuries). Without doubt it has received more blame than it deserved. If the shifting of that Papuan warm pool, however coincident with El Niño times, does not normally lead to drought, is it reasonable to point a finger at El Niño when it does do so? Off the west coast of North America a huge triangle of water, reaching out to Hawaii, was much warmer than normal, up to six degrees C. higher in places, in 1997. El Niño was pronounced guilty until it was realised that the warming had occurred before the Southern Oscillation could have had effect. Heavy winter rains in California were also blamed on El Niño even though previous El Niños usually made California dry. (There could be an analogy here with the penchant for blaming disease – Spanish flu, and the like – upon another place, it being so handy to indict somewhere else for all affliction.)

El Niño and its effects tend to be vilified whether or not they bring harm. Perhaps Californians welcome those heavy winter rains or, conversely, they relish the dry spells. Both forms of change cannot be resented, at least not by everyone, unless every variation from an average situation is begrudged. 'El Niño tightens its grip on the Galápagos,' stated a recent headline. The sea water was five degrees warmer, corals had turned white, and hammerhead sharks had vanished. What also happened there is that other species, those preferring greater warmth, arrived to fill the vacancies. On those all-important islands the greater rainfall of an El Niño time causes the vegetation to be much more abundant, and therefore a greater feast for the giant tortoises. Annual rings in their shells confirm this point by showing increased growth in tune with El Niño years. Therefore it is too bad about the hammerheads. It is too bad also for the marine iguanas whose seaweed food has been replaced by an inedible variety. But it is good for the land-based creatures, thanks to all the rain.

El Niño's fame makes it easy to forget that many other such fluctuations are also at work around the globe, even if less extreme in their effects. The NAO, the North Atlantic Oscillation, has been defined as 'one of the most dominant modes of global climate variability' after El Niño. It certainly swung widely between the winter of 1994–5 and that of 1995–6, with the latter year being colder than normal by two to four degrees C. all the way from Scandinavia to the Black Sea. Rainfall in much of that area was 50% below normal, whereas it was much higher than usual over the western Mediterranean and North Africa. The earlier year typified the conditions which had been in place, more or less, since the mid-1970s. What had happened was a swing from NAO's positive phase, with a low over Iceland and a high over the Azores, to its negative phase with much weaker influence emanating from the Atlantic. Hence the reduction in rainfall and the drop in temperature. It is almost needless to say, as did a report in March 1999, that 'the mechanisms behind (this oscillation) are still obscure'.

As yet no one is attributing to NAO the kind of global influence so blatantly wielded by the several El Niños which the world has recently experienced. It is this latest intensity, as well as the increasing frequency, that inevitably poses the question: will the present ENSO level be maintained or even worsened? Such a query also implicates human activity. Has the rise in greenhouse gases affected the ENSO system, and might it continue to do so yet more markedly? A team working at the Max-Planck Institute for Meteorology, Hamburg, reported in an April 1999 issue of *Nature* that gas emissions caused by humans will be influential in three directions. First, the average situation in the tropical Pacific may become more like the El Niño situation. Second, the year-to-year variations may become more extreme. Third, the variability may become more awry.

One difficulty with such forecasting is that the factors initiating and maintaining ENSO years are also not well understood. Neither are the effects of greenhouse gases, such as the cooling caused by increased cloud cover, itself induced by warming. Therefore the procedure of amalgamating one unknown with another unknown seems fraught with uncertainty, but the German computer models have been giving consistent results affirming that recent trends will probably continue. It therefore seems likely that ENSO, El Niño and La Niña will become yet more famous in the future, and also much more infamous.

By July 1999, when statistics had been suitably collated, Kelly Sponberg, from NOAA's Office of Global Programs, was writing to *Nature* that 'another El Niño "event of the century" had come and gone', with a 'wide variety of estimates of how much damage the world incurred'. These ranged from $14 billion to $69 billion, with damage from all natural disasters in 1998 assessed at $93 billion. Sponberg resented the 'lack of accuracy' as well as the 'general tendency to describe impacts through a few global totals'. There were also biases in the reporting, such as 'overestimating losses from industrialized countries and underestimating losses in developing countries or in areas remote from centers of government and mass media'. It was a good point he had to make.

A shack is as important to a shack-dweller as is a house in Miami to its owner, but their dollar values do not compare. On occasion the loss of a shack, with all one family's possessions, may be even more important than the loss of a house whose structures and valuables are swiftly replaced via insurance funds. Deaths in urban areas are recorded precisely, but elsewhere – as newspapers blandly state – can be 'several thousand' or merely 'numerous'. A tornado in Oklahoma hits headlines worldwide and inevitably influences thinking more than some similar damage in Bangladesh 'believed' to have killed ten times as many people. A flood in northern Nigeria once caused no immediate deaths and was therefore unimpressive as an incident, but it meant a closure of the only railway line for eight months. The damage done by that severance to the economy was intense, prolonged and also lethal in its way, but not sensational. A hospital in some third world area can quietly consume its funds, slowly empty its shelves of all medication, increasingly witness the departure of its nursing staff and eventually shut up shop. Such a happening is piecemeal and unexciting, without the gripping televisual drama of some natural furore, with dead, dying, trapped and injured individuals, such as wide-eyed children, by the dozens, by the hundreds, by even more.

These days an incident not seen on television is almost deemed not to have occurred. Such unreported happenings create less demand for action, less concern, less appeal for funds and aid. Small earthquake in Chile and few dead is still a recognised response, even in Chile if far from cameras. Kelly Sponberg concluded his letter: 'Not all impacts associated with climate variability are disastrous or even negative . . . (and) the popular vocabulary used to describe and record impacts of climate variability must move beyond deaths and dollar losses.' Some of El Niño's effects may have been advantageous. A volcano does spew forth material which, when weathered, is great for agriculture. Ill winds do tend to blow somebody some good, possibly even the majority. The word event, so often and so blandly used by scientists when referring to some ghastly natural happening, is possibly more suitable a term than disaster, mayhem, chaos, horror or catastrophe, the very words which, it must be admitted, have peppered many pages in this book.

Perhaps, when the world gets to grips with El Niño, and when the reasons for its arrival are as fully understood as all its consequences, it will be seen more as blessing than as curse. Expect a welcome deluge. Do not bother to plant seed if no rain will come. Realise that weather can vary, often tremendously. And keep in tune with terrestrial cycles rather than fight against them. We may be influencing these rhythms, but we are not in charge of Earth's climatic alterations, not by a very long way.

As an addendum to current thinking about El Niño, this little Spanish name having invaded modern minds so intensively, it is good to be reminded that the oscillation has an ancient history. The abrupt shift from cold to warm water off South America had worldwide influence long before the world began to realise that a bit of bad luck for Peruvian fishermen also brought a lot of bad luck for countless other people. Richard H. Grove, of the Australian National University, Canberra, has plainly done a lot of archival digging. Apparently there were bad ENSO years from 1685 to 1688, and also from 1877 to 1879, but there was an outstandingly difficult time between 1789 and 1793. Schooldays only told us that this was the anarchic period of the French revolution, with its best of times and its worst of times, but Grove has unearthed the bad news, in particular, from numerous different regions.

The governor of St Helena, knowing of 1791 droughts in India, in Montserrat and on his island, suggested these were all part of a single connected phenomenon. The South Asian monsoon had failed in 1789 and in 1792, but worst of all in 1790. Half the population were then to die in the northern Madras Presidency. These droughts were partnered by some damaging rainfall, with three Madras days in October 1791 witnessing 25.5 inches, 'more than . . . has been known within the memory of man'. Drought also occurred in Mexico, southern Africa, Java and Australia. Sydney's governor, recently arrived at a colony not initiated until 1788, considered the drought had begun in July 1790 and he recorded no rain until August 1791. The River Nile, whose flow and level have been measured for centuries, was extremely low between 1790 and 1797. Drought in Natal and Zululand between 1789 and 1799 was the most severe until 1862. The Antilles experienced a lack of rain not known since 1700. As for St Helena its drought, so disturbing and so thought-provoking to the island's governor, had started later than in the Caribbean, and only ended in mid-1794. Up in western Europe there was a drought in 1788 which followed a severe winter and a late, wet spring.

Perhaps the outbreak of revolution in Paris one year later was not an irrelevant coincidence after all. Historians provide many reasons for its arrival, but it may well be that El Niño added its mite of influence. Quite possibly all of history could be reviewed if every other ENSO season was as potent, and as well recorded, as the one which began, with such global intensity, in that deadly year of 1789.

Chapter 12

CLIMATE HISTORY

Planetary alteration – Gondwanaland and Pangaea – continental shift and drift – Chicxulub – Tertiary and Quaternary – glacials and interglacials – Atlantic conveyor belt – AD 535

Planet Earth, with its continents and oceans, mountains and deserts, cold places and warm places, appears very much set in its ways. It experiences hurricanes and earthquakes, and volcanoes and blizzards, but all such transient episodes soon vanish to reveal, once again, the familiar outlines of landscape and seascape which its inhabitants know so well. Such changes as are occurring – the Red Sea widening by one inch every year, Everest growing with similar haste, and temperature rising by less than one degree in the 100 polluting years of the 20th century – can seem so minuscule they lend support to personal belief that nothing truly changes. Besides, with each human's adult life span being shorter than 3,000 weeks, the observed changes are visually insignificant in any individual's time on Earth. Geologists consider that Red Sea widening to be 'extraordinarily rapid', and so it is in their epochal timescale, but its width differs by less than six feet from a human's first breath to his last. The distance from shore to shore therefore alters from an average of 200 miles to 200 miles and 6 feet during the whole of a human span.

This kind of arithmetic shifts dramatically when millions are introduced, with geologists casually using such numbers during their daily talk. One million increments of one inch, as with the Red Sea's 'speedy' increase, is almost 16 miles. One hundred million is therefore 1,600 miles, approximately the distance between the extremity of western Africa and the eastward bulge of South America. As Africa and South America were once united, with that eastern South American bulge fitting neatly into the Gulf of Guinea, the Atlantic only had to widen with Red Sea haste to form the current situation which seems to us so permanent, so inflexible, so fixed a feature of planet Earth.

In fact the map of Earth has never previously been exactly the same as it is today, with its particular landmasses, its well-known mountain ranges, its named lakes and oceans, and its several frozen zones. To begin with it was not remotely like the modern situation, with the portions of land wholly different. Rodinia, the earliest known supercontinent which existed 650 million years ago, had a shape bearing no resemblance whatsoever to today's outlines. Neither did Gondwanaland and Pannotia of 500 million years ago. One hundred million years later Laurentia, or North America to be, was colliding with Baltica of northern Europe, causing the Appalachians of New England and the Caledonians of Scotland to arise from the squeezing down below, but still the dry land all around them looked nothing like the modern shapes. Some 300 million years before now there was one enormous continent, Pangaea, stretching from pole to pole.

Then, with the landmasses all united, they started to fragment, and by 100 million years ago the beginnings of the current outlines can be discerned. North and South America were still separated, remaining so until 3.5 million years ago, and India was still remote from Asia. Then that Indian wedge of land started pushing northwards from its southerly location, not only forcing up the Himalayas but linking with Asian territory some 50 million years ago. At the same time Australia and Antarctica were dividing to go their separate ways, with Antarctica remaining in the south and Australia heading north. The map of continents was then recognisable, and akin to modern maps, even if a long way from identical with today's configuration.

As for ice, and as for temperature in general, some students of Earth history believe that Earth's oceans may *all* have been frozen several times between 750 and 570 million years ago. According to a *Nature* article in December 1998, sent from Pennsylvania State University, the Earth used to lean in such a way that its tropics of the central equatorial belt were colder than its poles. It is even thought that the build-up of ice may have been sufficient to tip the Earth towards a flatter angle to the plane of its orbit than its current steeper inclination of 66.5 degrees. Such a massive quantity of ice may seem remarkable, and hard to accept with the sun just as strong as it is today, but the process of cooling can be self-inducing and, once started, will continue. As soon as there is a sufficiency of ice and snow this brilliant and reflective covering will bounce back so much solar radiation that further cooling becomes certain. Similarly, and contrarily, any melting not only causes more land and more water to be exposed but more solar heat then to be absorbed. Such a balancing act between hot and cold must always have existed, swinging one way or the other. Temperatures are thought to have reached a maximum 55 million years ago, round about the time that India was reaching its current destination. Cooling then occurred for 20 million years, so much so that Antarctica became ice-covered, this continent staying that way.

Along with the landmasses being extremely different long ago so was the atmosphere. To begin with, when life was starting to evolve, the gas around the Earth consisted of nitrogen, water, carbon dioxide and carbon monoxide, but it lacked oxygen. Only when certain bacteria had evolved a system akin to botanical photosynthesis did their principal waste product, oxygen, become added to the atmosphere. Its level has fluctuated over the billions of years, sometimes rising as high as 35%, and it is currently 21%, the proportion we enjoy. This gas was therefore a pollutant in its earliest days and is now a prerequisite of (almost) every living form.

Hastening on to more modern times, when there was certainly a modern atmosphere, ice began to form on Greenland about seven million years ago, a northern lowering of temperature encouraged, it is thought, by Africa's surge northward. This continental push gradually enclosed the Mediterranean and halted a warm water flow from the inland sea into the Atlantic. Instead of that flow outwards there was a far smaller flow of ocean water into the Mediterranean. With evaporation there proceeding apace, and with its water boosted only very slightly by the surrounding rivers, the Mediterranean became extremely saline and it eventually dried up. Meanwhile, and in converse fashion, the nearby Atlantic Ocean became less salty. Consequently its waters were more likely to freeze, notably in the north, and the more they froze the more light and heat was reflected back to space. An ice age was therefore more likely to occur.

With continental drift steadily forcing Africa northwards, and with this movement altering the geography of Morocco and Gibraltar, there came a time when a breach in the dam between Spain and Morocco became more probable. The low-lying Mediterranean region, by then a desolate landscape similar to but far larger than California's Death Valley, was all ready for the filling. According to Kenneth Hsü, of the Lamont-Doherty Earth Observatory in New York, Atlantic Ocean water was able to cut a tiny channel in that Gibraltar dam. A torrent of it then swept through this miniature opening, causing the 'Gibraltar Falls [to become] one hundred times bigger than Victoria Falls and a thousand times grander than Niagara'. Later, and when the earlier situation had been restored, warm water flowed once again from the Mediterranean into the Atlantic. Hsü believes that this 'salinity crisis' lasted for 400,000 years ending about 5.4

million years ago. It is also thought that both an emptying and a refilling of the Mediterranean may have happened several times in its history. Therefore the nearby Atlantic was extremely influenced, one way or the other, with its waters becoming either saltier or less so. Water's salinity also influences its density, as well as its propensity to form ice when cold, and therefore affects its rate of sink. (Ocean water/fresh water weights are in the proportion of 64/62.5. Even humans can notice this difference, it being much easier to float in sea water than in ordinary swimming pools. As for the Dead Sea its extreme saltiness makes floating a most worthwhile activity.)

The spate of recent ice ages started 2.5 million years ago when Greenland's frozen surfacing spread south, reaching deep into the United States and submerging the British Isles as far south, more or less, as the River Thames. With so much water then turned into land-based ice not only were sea levels lowered, by several hundred feet, but there was much less precipitation. In short, and drawing a line after this gallop through much of Earth's history, alteration has been its keynote. There has always been chop and change, not just very long ago but also much more recently.

Biological evolution is generally portrayed as a smooth advance, with single-celled organisms becoming multicellular, with invertebrates fanning out in innumerable directions, with vertebrates then appearing as further major difference before these backboned creatures diversified into amphibia, into reptiles, and finally into birds and mammals. Botanical life altered in similar style from simple to more complex, from soft to woody stems, and in time to all the effulgence of flowering plants. Such steady change might make one assume a prevailing stability in the earthly conditions partnering all this evolution; but this was not so, most definitely not so.

There was always alteration. The continents were moving. Temperatures were changing dramatically. Land was becoming water and then land once more, time and time again. Ice sheets came and went. Precipitation varied from generous to nothing of the kind. Islands were formed and then submerged. Mountains arose before being eroded to become mere stumps of their former selves. There was perpetual variation, accompanying and encouraging the unceasing permutations of all the living forms.

In addition there were cataclysms, not merely of the local kind, as volcanoes blew their top or earthquakes cracked the land in two, but of utter devastation. The geological record indicates a dramatic shift between the Permian and the Triassic some 250 million years ago. Almost every form of life then died – with some palaeontologists affirming the extinction of 96% of all species of plants and animals. This period of destruction took place within a few hundred thousand years, both on the land and in the sea, and soaring temperatures are thought to have been responsible. Pangaea, then the single continent, was coupled with a single ocean which stretched for 220 longitude degrees, as against today's Pacific which embraces only 130. Whatever caused this awesome upset is uncertain, and heat may not have been involved, but there is no doubt about the mass extinction.

Such an abrupt species loss between those two ancient ages of the Permian and the Triassic was the greatest of its kind, but there were others. Between the Triassic and the Jurassic, 200 million years ago, there was a further species disappearance. So too between the Jurassic and the Cretaceous 150 million years ago, and then, most famously if not most destructively, between the Cretaceous and the Tertiary a mere 65 million years before the present. That was the occasion, probably induced by some meteoritic impact, which not only extinguished the dinosaurs but countless other forms of reptile and all manner of different species.

This mass extinction meant the disappearance in the fossil record of over 40% of fossilisable life forms and, so it has been assessed, the vanishing of 75% of species in existence at the time. Both Americas were hardest hit in terms of species loss, a further indication that Mexico's Chicxulub region received the impact which is generally believed to have caused the devastation. Most species extinction does not occur in such dramatic fashion, being more gradual and piecemeal, but the massive slaughterings throughout Earth's history are disproportionately important because they tend to remove the resilient and incumbent groups. Dinosaurs were such a group, lording it over the land – and sky and sea. With their disappearance the small and, until then, insignificant mammal species could take advantage of the altered situation and flourish accordingly – on land and sky and sea.

A mere ten million years after the dinosaur extinction, and therefore 55 million years ago, there was a further massive killing, primarily of creatures living on the seabed. Some 50–65% of all such bottom-dwellers perished within 1,000 years, their disappearance coincidental with deep water warming from 11–15 degrees C. The cause of this disturbance is unknown, but a formidable release of methane, a powerful greenhouse gas, is believed to have been responsible. There is argument that layers of methane hydrate, created by bacteria from organic material, suddenly erupted, or 'belched', to cause a speedy increase in atmospheric methane levels. These, in their turn, led to a global temperature increase, via greenhouse warming, of five to six degrees C. Such a surge in temperature was sufficient to kill off countless species, much as the meteorite had done ten million years earlier.

The next major climatic shift occurred between the end of the Oligocene and the start of the Miocene, some 27 to 21 million years ago, with several intermittent periods of a cooler climate, these being most manifest every 400,000 years or so. Ten million years later, and during the Miocene, temperatures were higher, despite low carbon dioxide levels of 180 to 290 parts per million. In short, whether or not these assessments of earlier conditions are 100% accurate, it is agreed by all investigators that Earth's climate has been extremely varied in its lengthy history. The animals and plants had to adjust accordingly, or fail and die.

Geologists do not identify the various epochs and ages arbitrarily, as if these are merely rungs upon a ladder, each conveniently spaced to serve as guidelines for the passing years. Instead the demarcations distinguish major alterations occurring between one period and the next, these being detectable in the various kinds of rock being laid down at the time. Planet Earth has not been constant, either in its temperature, or in its mix of land and water, or in its surface properties, or in any characteristic which can be identified and measured. Life, in all its myriad forms, has therefore had to alter and become in part extinct while being replaced by a sufficiency of other forms better able to survive. It is almost a greater wonder, bearing all the change in mind, that life in any form has managed to remain.

Earth itself, as a revolving and orbiting planet, has also been exercising its own and personal variability, much as any planet does. In the distant past Earth's elliptical orbit, its angle of tilt and its gyrations were all dissimilar to the current situation, but even today's apparent stability is steadily altering in a rhythmic manner. The basic cycles, of eccentricity, obliquity and precession, each have different periods and oscillations.

1. Earth's orbit around the sun, an ellipse rather than a circle, steadily changes its eccentricity, namely the ratio between the greatest and smallest diameters of that ellipse. Currently, Earth's ellipse brings our planet to within 92 million miles (at its closest) and 95 million (at its furthest) from the sun. This ratio of eccentricity has a cycle of its own which oscillates every 100,000 years or so.
2. 'The planet Earth rotates on its axis once every 24 hours' is a statement drilled into schoolchildren, leaving bewilderment whether it could rotate on anything else (and where that axis might be, and whether

footballs have one). Currently this line of tilt is canted at 66.5 degrees to the plane of Earth's orbit around the sun, the fundamental fact which causes our summers and winters when one or other hemisphere is receiving a greater share of sunshine. This angle (its obliquity) may seem constant, but actually varies between 21.6 and 24.5 degrees within periods of 41,000 years.

3. The times of perihelion and aphelion in each year also vary but with a period of 23,000 years. The dates of closest and furthest distances from the sun may therefore be 17 February and 17 August, or 24 September and 24 March, or any other six-month pairing. It so happens that today's situation, early January and early July, means that northern winters are not as cold as they would be if these datings were reversed, and if Earth's greatest distance from the sun occurred during the northern winter.

Just as the climate of one particular day, with its winds, rainfall and insolation, never precisely equals the climate of any other day, so has the present configuration of 2 January and 2 July, of 23.5 degrees of tilt, and the current nature of Earth's ellipse, never been precisely matched in some earlier year. Those three distinct cycles of 23,000 years, of 41,000 years and of 100,000 years can never have coalesced in a fashion identical to the present arrangement.

Therefore if the three forms of dissimilarity are coupled with shifting continents, with shifting ratios of land to ocean, with changing quantities of ice or cloud, and with altering heights of land massifs plus the general oceanic level, it is more astonishing that any form of climatic stability has ever occurred. Meteoritic impact, volcanic effusion and overwhelming storm are further extras, adding to the degree of change, sometimes modestly or even gently, but occasionally horrendously. The sun's output may have been similar from one year to the next, from one millennium to its successor, and throughout much of Earth's several billion years of history, but the manner in which that radiation has been received has altered with every ground-based alteration. Consequently, and throughout our planet's relentless turbulence, the innumerable forms of life have had to continue with the replication of living matter in whatever forms proved most satisfactory (or least unsatisfactory).

Climatic change becomes more intriguing, and slightly easier to discern, the closer it has occurred to modern times. The four divisions of the Tertiary Age, which began about 69 million years ago, are Eocene,

Oligocene, Miocene and Pliocene. The Miocene and Pliocene ('less new' and 'more new' in Greek) were the last two epochs, these coming after the Eocene ('new dawn') and Oligocene ('few new', with little to record). The Miocene lasted for 19 million years and the Pliocene for five million. These two were followed by the Pleistocene ('most new'), this being the first epoch of the Quaternary Age. It lasted for most of the recent two million years. As for the present time, according to geological naming, we are existing in the Holocene ('wholly new') epoch, the second such epoch of the fourth age. (There is a move to rename Tertiary and Quaternary as Palaeogene and Neogene, but the older names are being tenacious, certainly in this book.)

The four most recent epochs from Miocene to Holocene have witnessed considerable climatic change, an above-average quantity of alteration despite their paltry timescale relative to the billions of years of life on Earth. For example, a major extinction of mammals occurred during the late Miocene in North America, including 9 of the 18 species of horse. The reason for their vanishing is unknown, with one theory being the growth of prairie grass, itself caused by different conditions which proved difficult for animals with shorter teeth. During the even later Miocene the closure of the open seaway between North and South America, open for almost 100 million years, caused a massive readjustment of oceanic circulation, this beginning some 4.6 million years ago. The Gulf Stream was thereby intensified, or even initiated, bringing warmer waters northward and also increasing deep water currents bringing cooler water south again. This circular flow of water is said to equal the flow of all the rivers in the world – or 80 times that of the Amazon. The isthmus of Panamanian land between the north and south Americas may have been one reason, out of a bewilderment of reasons, why the Miocene was followed – roughly 3.0 to 2.6 million years ago – by a time of progressive global cooling, this leading to large-scale terrestrial ice sheets in the northern hemisphere.

Each time of cooling was followed by a time of warming before another cooling came along. This seesawing between warmth and cold has been a dominant feature of the Quaternary Age and did not occur in similar fashion during the Tertiary times. There have been dramatically cold periods during Earth's lengthy history, such as the dominating cold of 600 million years ago, but not the warm-cool-warm-cool on-off fluctuations of the past two million years.

No one knows why this should suddenly have been so, but of course there are theories. One of them implicates that Central American closure, with this sudden blockage surely being disruptive to the traditional ocean currents. Another is linked to the shifting landmasses which caused the Arctic Ocean to become even more landlocked. The Bering Strait and the Greenland–Norway gap are still open, but both openings are thought to be narrower than ever before. None of this change happened overnight, but – so it is argued – a situation was reached when ice ages became more probable and that probability then occurred, perhaps quite speedily.

Or perhaps Earth's own orbit abruptly became more relevant. The quantity of heat which Earth receives from the sun at its times of greatest proximity is more than 1% greater than at times of greatest distance. Perhaps this modest difference had become significant. If so, and if it did form a major influence, one immediate curiosity is why such variation did not cause a plethora of ice ages throughout the Tertiary. Perhaps it was inadequate on its own, and maybe it needed to act in concert with other changes – that Panama closure, and that greater Arctic confinement, for example – to be so critical. A final straw, to use this refrain yet again, *does* break a camel's back. A final shift of land, a little extra heat, a changing current, a changing anything can, seemingly abruptly, cause quite a different circumstance to occur.

With the arrival of humankind some of the changes are thought to have been induced, or at least accelerated, by this new form of creature, a species more capable than any other of altering its environment. A survey of carbon abundance during the past million years, recorded in marine sediments off the coast of Sierra Leone, indicates a sudden increase of this element about 400,000 years ago. Such an accretion is presumed to have arisen from vegetation fires which were possibly man-encouraged or even man-begun. In later centuries, as was also shown in the same sediments, there were further carbon peaks when global climate was changing from non-ice age back again to ice age. Our early human ancestors may not have been responsible, but such individuals were then in existence, their use of fire is well documented, and land clearance or game-hunting by fire both have ancient histories.

Drilling deep within the ice in both polar regions has been extremely revealing, with the work itself excitingly advanced in recent years. Soviet scientists started drilling from their Vostok base in eastern Antarctica during 1980, but this work had to be abandoned in 1985 after the drillers had reached 7,224 feet. A second borehole, started in 1984, went down to 8,353 feet after six years. (The Soviet Union used to have eight Antarctic bases, but only four still remain today, including Vostok.) Other nations have also taken part in drilling operations, with the Japanese achieving 8,211 feet in 1999 beneath Dome Fuji, East Antarctica. Similar work by the United States in Greenland struck bedrock at 9,934 feet and at 10,016 in two different locations.

The upshot from all this deep-ice drilling is that glacial and interglacial times can readily be identified, just as tree rings in timber have much to say about the climate which prevailed when the rings were made. Ice cores also indicate the quantities of methane and carbon dioxide present in earlier atmosphere, as some of this ancient air became trapped within the falling snow when each layer of ice was being laid down. These two most important of the greenhouse gases, aside from water vapour, have a lot to say about climate in general. The cores indicate their previous levels, and therefore provide telltale portraits of the four recent transitions from glacial to warm. These started about 335,000, 245,000, 135,000, and 18,000 years before now. As the intervals are, more or less, 100,000 years apart it is tempting to associate them with the 100,000-year cycles, listed in 1. above, involving Earth's elliptical orbit around the sun.

The ice cores and ocean sediments also provide information, notably by examination of oxygen and its isotopes, of both the general global temperature and the actual quantity of ice present in the polar regions of each hemisphere. At the peak of the last glacial event, some 20,000 years ago, there was so much water locked up in the massive volumes of polar ice, these reaching down, it should be remembered, to the 45th parallel in North America, that the world's sea level was 400 feet lower than it is today. The continents therefore had quite different outlines. At that time *Homo sapiens* was very much in evidence, physically identical – it is assumed – to modern humans, drawing or about to draw in caves, and soon to embark on the Neolithic revolution which led to agriculture, to domestication of livestock, to large-scale settlements – and to our modern world. As the ice melted there was a speedy rise in sea level and all sorts of geographical alterations were then occurring, such as the creation of the English Channel, that severance thought to have taken place about 7,000 years ago.

It is easy to suspect that this channel's opening, the transformation of Britain into an island, happened very rapidly. There would not have been a Red Sea widening, mere inches every year, but a sudden breaching of the slender neck of land between what are now southern England and northern France. The narrow opening which resulted would then have been battered from both sides, with tidal water rushing into and then out of both the Atlantic Ocean and North Sea. Instead of dry land between the two north-western bits of Europe there would, in all probability, have suddenly been a gap of hundreds of yards, these hundreds becoming wider very speedily, much as happens when any dam gives way. Unfortunately all evidence of this transformation has been obliterated, with the waters totally triumphant and destructive in their assault. They are still celebrating, with the Channel widening every year.

But there is exciting evidence in North America where a similar event occurred. As the ice began to melt in that region 15,000 years ago it created great lakes of melt-water, such as the one – known as Lake Missoula – filling several valleys in Idaho and western Montana. This is believed to have collected 600 cubic miles of water, the combined equivalent of lakes Erie and Ontario. Then, one day, this accumulation broke through the surrounding ice still serving as barrier. This wall was not of concrete, built to resist such cracking, but an amalgam of ice and earth only too ready to yield when pressure upon it became overbearing. The released water then surged into the Spokane valley, headed west, paused awhile in what is now known as the Channelled Scablands, and soon broke through a further blockage to encounter the Columbia Gorge and, eventually, the sea. It is thought, by those who have studied the visible evidence of carved valleys and a seared landscape, that the whole event only lasted a few days. During that time its flow possibly equalled that of all the rivers in the world. What a sight that would have been for any bystander (on suitably high ground), for there were people in North America at that time.

As for planet Earth's four recent ice ages, and the changes in temperature from glacial to interglacial so clearly demonstrated by the ice cores, these transitions were each accompanied by increases in atmospheric carbon dioxide. The proportions changed from about 180 parts per million by volume to 280–300 ppmv. At the same time the quantities of methane increased from 320–50 parts per billion by volume to 650–770 ppbv when the four glacial epochs were altering to become times of greater warmth.

It is highly pertinent, not to say frightening, that the present concentrations of these two major greenhouse gases now exceed the highest concentrations in earlier times when glacial epochs were becoming warm, and when the so-called interglacials were either on their way or had arrived. There has never been a time in the last 400,000 years when, according to all those cores drilled through the ice, levels of CO_2 and CH_4 were as high as they are today. The highest CO_2 level in the past 4,000 centuries occurred 300,000 years ago when there was a lower concentration of the gas than there is today. The ice ages came and went, always in tune with the greenhouse gases whose levels also rose and fell.

Those ice ages did alter the landscape quite amazingly. Think of most of Britain submerged beneath hundreds of feet of ice. Think of the northern United States being similarly interred, and do not forget the chilling climate which partnered such frigid burial. The arrival of an ice age would not mean a slight adjustment to life but, for those north either of the Thames or of Colorado and Missouri, it would mean one's previous homeland becoming submerged beneath hundreds of feet of ice. It would also mean, for everyone who had retreated southwards just beyond the ice, having the kind of climate now experienced by northern Canadians and northern Russians. The soil is permafrost. Trees do not grow. Wild animals are scarce. The cold is paramount, and the massive wall of ice is just a little further north.

Planet Earth, plus all its citizens, is now enjoying an interglacial, and has been doing so for some 11,000 years. Therefore the man-made and unnatural effusions of carbon dioxide and methane are adding to an already high, but naturally high, situation. The present global concentration of carbon dioxide is 365 parts per million by volume and of methane is 1,700 parts per billion by volume, with that CO_2 quantity slightly greater than any of the earlier natural peaks. The current amount of CH_4 is greater still, being over twice as high as has ever happened during the past 400,000 years. Everyone expects today's abnormally high levels, so encouraged by human activity, to reach even higher concentrations in the immediate future. We may tinker with the present concentrations, achieving more efficient cars, more nuclear power, less deforestation, more regard to the environment and less squandering of resources, but the six billion of us will not mend our ways sufficiently or in time to stop those levels rising further still. Besides, another five billion people are on their way to join our existing multitude.

It is therefore no wonder, with Earth experiencing a period of interglacial warmth following the retreat of the last ice age, that the advent of yet more greenhouse gas is of concern. We are not warming ourselves from a time of cold to one of warmth, but from one of warmth to something hotter still. The anxiety about global warming is therefore more than justified. It should be shouted from the hills. Or, say cautious sceptics who have opposing views, such worry should be shouted with reservation, like people who only think their house is burning. The issues, in short, are not clear-cut.

During the 1960s, particularly when a couple of severe winters, such as 1963, had hit the northern hemisphere (always more thought-provoking for northerners than anything down south) there was even talk that a 'little ice age' was on its way. It was believed this might be akin to the icy European winters of the 16th and 17th centuries when the Thames was frozen, when oxen were roasted on its ice, and when every other Flemish painting seemed to be of individuals skating amid a landscape of widespread snow, beset with stationary windmills by solidified canals. There may, of course, have been some artistic licence about those times, with snowscapes perhaps more marketable than ordinary land. Was there truly so much cold? As for the Thames what did constitute a freezing? Was it solid right across, obliterating the ferry trade? And was it always oxen, so much better for a story than lesser animals? The last 'genuine' freezing is thought to have occurred in 1683 when Charles II had two more years ahead of him.

By the 1970s, when technology was vastly different and after some milder northern Januarys, icy thoughts became generally replaced by the possibility of greater warmth. Better computers were able to provide better models of more likely outcomes, and by the 1980s most thinking had coalesced to create widespread talk of global warming; but there was still no confirmation that humanity was to blame. Neither was there universal belief that climate forecasts concerning a decade, or two, or more in the future were likely to be spot on.

As every human knows who has ever planned a picnic or failed to insure a fête it is possible for weather predictors to be a trifle misleading, or even horribly wrong. In 1889 Jerome K. Jerome was already mocking them in his classic *Three Men in a Boat*, written only 11 years after official forecasts had begun. The first weekly weather report was published by the British Meteorological Office on 11 February 1878. Now, despite weather satellites, a global network of recorders, powerful computers and far more knowledge, doubt can still exist about meteorological ability to foretell the future. The actions and interactions of this wind, that cold front, those clouds and this precipitation are so enmeshed that weather predictions further than five days ahead are fraught with potential error. It is therefore inevitable that scepticism will arise concerning climatic prognoses not merely for the next five days but about the planet's health during the 21st century. However there is a world of difference between assessing whether it will rain in some locality at the weekend and understanding that greenhouse gases will raise the temperature. The short answer is that such gases do raise the temperature. It should therefore be no surprise to hear meteorologists affirming that global warming is not only here to stay but will assuredly increase.

At a conference in Utah during 1998, organised by the American

Geophysical Union, there was a disquieting revelation. It had already been accepted that there could be planetary temperature swings of up to ten degrees Centigrade, these leading to, and associated with, glacials and interglacials, but at Utah it was being proposed that such a changeover could take place in as little as 20 years. Subsequent discussion therefore embraced the possible causes of such speedy alteration. Ice sheets do melt, and glaciers do retreat, particularly the unstable ones, but these differences are thought to be responses to climate change rather than its cause. Solar luminosity may also vary, and does so detectably, but past solar warmth cannot easily be measured. Therefore, as a possible cause of the dramatic shift, the sun must be set aside.

The finger now points towards ocean circulation. The 'most widely recognised hypothesis', according to a reporter of the Utah conference, is that deep water in the North Atlantic 'flips' between two stable modes of operation. A seesaw, as analogy, has difficulty in staying horizontal. Any small weight will tip it in one direction, and a slightly greater weight will not only tip it back again but even past the horizontal. The seesaw device is inherently unstable, and ever ready to be upset.

Similarly the surge of Gulf Stream water proceeding northwards off the eastern shoreline of North America loses its heat in the northern latitudes, much to the benefit of western Europe, and then sinks deep to travel south again. The more heat it delivers to northern areas the greater and faster will be that degree of sink. Equally true is the fact that the greater the sink, possibly encouraged by more ice in northern regions, the greater will be the flow of warm water northwards from the south. Or maybe the water's freshness, as distinct from saltiness, is also in part to blame. More rain or more snow will cause the water on which it lands to change its weight, and oceanographers have been detecting differences in salinity which may, in turn, be helping to alter the global water cycle. As with slight changes in temperature having a major effect so may alterations at different levels in an ocean's saltiness be markedly influential.

The oceanic cycle of the Atlantic, known to its investigators as the conveyor belt, is thought to be sufficiently adjustable, perhaps by minor alterations, for it to flip from one mode to another. If there is less sink there will be less warm water from the south, and if there is less warm water there will be less sink. There is a neatness to this notion, save that its detractors have raised some awkward questions. How come that the Atlantic oscillation is so in tune, and similarly timed, to distant oscillations, as in the seas off California, for example, or around Indonesia? Why have methane levels in the atmosphere, presumably caused by microbial activity on tropical land, risen in keeping with the oscillations? And why have Arctic and Antarctic temperature levels sometimes varied inversely with each other? As a further major point, is not the Pacific Ocean just as influential on global temperatures, if not more so, and why should a stable Pacific situation suddenly alter for no apparent reason?

All such questions are entirely valid for the global debate, even if they are as yet unanswerable, but they do reinforce existing awareness that climate is a multitude of different happenings. Climate change, despite the simplicity of that on-off seesaw analogy, may result from a plethora of different happenings, each adding its mite of alteration until there is sufficient influence to effect a dramatic transformation. In short how can all these interactions be so defined and understood that their final united influence is known – for sure? And how can the very cleverest of computers, presented with the very best of available information, possibly get it right?

One overriding obstacle, still not unravelled, is how to date and coalesce the records from southern and northern hemispheres. Was this particular blip on this particular southern ice core laid down the very year, or decade or century or millennium, when a similar blip was being laid down in the north? If it occurred earlier it might have caused the northern event, but if later it could not have done. The synchronisation of such variables has to be known precisely, or far more precisely than at present, if causes and effects are to be better understood. Accurate timing of differing events lies at the base of all such work.

Everything becomes much easier when prehistory gives way to modern times, particularly after literacy had been established. Even so the deciphering of earlier climate is still not plain sailing, as Mike Baillie (already mentioned in Chapter 8) and David Keys will readily testify. They had observed a kink, a glitch, a hiccough in annual tree rings from a multitude of timber sources. Dendrochronology is the study of time via growth circles. Matching such rings from dead and desiccated trees, from wooden artefacts, and from growing timber can jointly provide a year-by-year record of good and bad times dating back 7,500 years. It is therefore possible to say that one particular set of rings, denoting a changing and distinct pattern of growth, happened not just during the first millennium, or the first century, but between one definite year and another definite year. David Keys observed that AD 535 and 536 were both very bad years for growth. The tree rings at that precise point in history were nothing like so broad or so well made as was customary.

Why should there have been such a hiatus? Fortunately there were literate individuals around the world at that time who described those years. Such official scribes did not normally record the weather, as if they were holiday-makers sending postcards home. Instead they concentrated on weightier matters involving religion, politics, legislation, wealth, royalty, or all five in unison with all five being so intimately involved, but the two years of 535 and 536 were so extraordinary that the inscribers broke with custom. John of Ephesus, writing in Constantinople, described how the sun was so dark he feared it would never recover. Cassiodorus penned much the same from Rome. So too a paintbrush writer in Japan who called it 'the year of dust'. As for Britain that was, more or less, the year King Arthur died, this death perhaps more symbolic of a darkened situation than an actual demise.

Such widespread agony, coupled with the solid evidence from trees, suggested powerfully that something dire had happened. To raise sufficient dust to disturb photosynthesis for at least two years indicated, according to various calculations, either terrestrial collision with a 2.5-mile diameter asteroid or a four-mile comet or the massive eruption of a volcano. Of these three possibilities the third became most favoured. No appropriate and asteroidal impact site had been discovered, and no writer had described a comet, with comets often mentioned as presumed portents of disaster. Furthermore the Tunguska event of south-eastern Siberia in 1908, described in Chapter 8, is generally believed to have been a comet's air burst, but it caused no darkening of the sun, not even locally.

Dusty evidence of the 535 happening has been found both north and south of the equator. Therefore the volcanic eruption, if a volcano was to blame, took place sufficiently near the equator for its dust to go both ways. There are 90 possible volcanoes suitably located, and a study of ancient Asian documents from that fateful year caused suspicion to fall upon Krakatau, it being in the right place to satisfy their various descriptions. Unfortunately, although charcoal has been removed from various levels of this volcano, and the samples have been radio-carbon dated, not one of them points to its creation during that all-important year of AD 535. Krakatau did pull off a similar explosive trick in 1883 – massive detonation, global darkening, worldwide dust, and this 19th-century event may have been equivalent to the occasion 1,348 years earlier; but, if all those early texts and trees are to be believed, the 1883 explosion was a lesser devastation than the earlier happening. The old texts and their inscribers perhaps tended towards exaggeration, but not those trees. They are perfectly objective and impartial in their blunt portrayal of an awesome time. (For more on this fascinating blend of botany, climatology, history and vulcanology, read *Catastrophe* by David Keys.)

History has provided even better records in the years since that Dark Age sixth century, but the cause of climatic change is rarely so blatant as with that Indonesian volcano. With regard to later years it is generally believed, from a multitude of sources such as vineyards in England's Yorkshire, that a warm period lasted from AD 1100 to 1300. A much colder time then followed, lasting from 1400 to 1850. How much colder it then was, when icicles hung by the wall and milk came frozen home in pail, is hard to know, but it was occasionally sufficient, as already mentioned, to freeze the River Thames. Alas, but the Thames flowed freely during the exceptionally cold winters of 1940, 1941 and 1947, presumably because its waters were then heated via association with power stations and industry in general. This extra warmth forbids a modern freezing, however much Londoners would undoubtedly relish such a carnival and all its roasted ox.

As for that so-called 'Little Ice Age', lasting from the 15th to the 19th century, this colder European spell did seem to have occurred also in the western hemisphere and in Earth's southern half. It was therefore widespread, but there is still no explanation for it. Perhaps the climate then was not so abnormal and exceptional. Perhaps the earlier mediaeval warmth and the current warmer time are the aberrant periods. And perhaps, with today's greater warmth having increased, although initially very slowly, as from 1850, the man-made greenhouse gases were even then starting to take effect. They were certainly forming a greater proportion of the atmosphere, notably CO_2. There were neither cars in the mid-19th century, nor aircraft nor other oil-consuming systems, but dark satanic industry was in full swing, with prosperity assessed by chimneys and by smoke.

The 20th century has not only been the best recorded century of all time, but it has ended with greater concern about the weather than had ever previously existed. For the century's first half, despite increasing industry, and increasing levels of greenhouse gases, and increasing people and increasing everything (well, try thinking of decreases), alarm bells did not sound about climate in general. Nor did they sound at the start of that century's second 50 years. Then came the 1980s, the warmest decade since proper records had begun. Average global surface temperature had been rising since the 1920s, with an increase of about 0.4 of a degree C. between 1920 and 1940. Temperature was then stable for a while, but it rose again at the start of the 1980s so that, by the late 1990s, it was more than 0.6 degrees higher than it had been at the start of the most recent century.

The 1990s experienced even warmer years than the 1980s, with 1995 breaking all earlier records. In total eight of the nine warmest years on record occurred in the 1980s and 1990s. Inevitably there was concern, particularly among climatologists who had long been observing the trend, but this preoccupation did not spread immediately. The IPCC, Intergovernmental Panel on Climate Change, only had its first meeting in November 1988. The panel's initial imperative, understandable in the circumstances, was to ask for a scientific report. This was published in May 1990, and the second major conference on world climate took place in Geneva later that same year. By the time of the United Nations Conference on Environment and Development, held in Rio de Janeiro in 1992, the matter of global warming had really taken hold. There had never before been such an international conference, not only so well attended by the world's leadership but also by 25,000 other delegates. The global subject had firmly arrived on the politically global agenda, and is now unlikely ever to depart. The people of this planet Earth have collectively realised that their single home needs attention. It is high time that they are doing so.

Global warmth and rising temperatures are now subjects so ripe for general discussion that they are eclipsing other climatic variations, not least the presence or absence of drought. Recent sediment studies from Lake Naivasha, Kenya, have shown great variability in water level and aridity. This lake tends to yo-yo even in a single human lifetime, but the new work (published in January 2000) has discovered that, although the lake never dried completely, there were marked periods of severe drought even within the past millennium, each of them more severe than anything witnessed in the 20th century. These happened from 1390 to 1420, from 1560 to 1625, and from 1760 to 1840, all three much more positive than the relatively minor changes to have occurred since measuring instruments have been in place.

There is much more human suffering among societies affected by declining or irregular water resources than is caused by shifts in temperature. Rises of one or even several degrees of warmth are influential, but the presence or absence of moisture makes living either possible or quite the opposite. In one such situation crops will grow and animals will drink. In the other they will perish, along with everyone dependent upon them. That total of 175 drought years between 1390 and 1840 must have made the region around Lake Naivasha, and probably far beyond, nothing like the game-rich savanna landscape which so enchanted the foreign railwaymen a century ago as they pioneered their steel track north-west from the site of modern Nairobi towards Uganda.

The Naivasha work ties in with local oral history about periods of famine, of political unrest and mass migration. It also, according to Frank Oldfield when writing about the African studies in *Nature*, shows 'how such research can help to provide more realistic projections of the effects of climate change in regions where human societies remain hostage to an uncertain water supply'. Shifting weather patterns will affect all of us, but the effects will be markedly more severe among people already living on the brink, when a slight lessening of rain or a shift in soil moisture is not a matter solely for interest or debate and, perhaps, a slight adjustment in livelihood but entirely crucial to life itself.

A terrible wrongness follows from the fact that, although most human influence upon climate change has been, and is being, caused by industralised societies, the results will be most critical for those other societies leading more traditional, less influential and, in general, less certain lives. The sins of the fathers are therefore falling upon entirely different families in entirely different portions of the globe.

Chapter 13

ATMOSPHERIC GASES

Ozone – UV radiation – Antarctic hole – Montreal protocol – CFCs – methyl bromide – carbon dioxide – greenhouse effect – emissions in general – Kyoto – carbon sink and sequestration – water vapour – methane – sulphur dioxide

The triple form of oxygen, with three atoms to each molecule instead of the conventional two, receives a mixed press. On the one hand it is more than welcome, helping to protect us and our skin from harmful ultraviolet radiation. On the other it is a smelly and unpleasant ingredient of urban pollution, damaging to vegetation and harmful to human health. 'Ozone levels rise disturbingly' can be one newspaper headline while 'Ozone levels fall disastrously' can be another, perhaps in the same paper and on the same page. The crucial difference lies in where this form of oxygen, first discovered by Christian Schönbein in 1839, actually exists. (That was shortly before this German chemist invented gun-cotton!)

Most of the atmospheric ozone is unevenly located in the stratosphere, some 7 to 30 miles above the Earth's surface. Even so there is not much, the quantity being equivalent to a layer around the Earth one-eighth of an inch thick. Most of this ozone is created over the tropics where sunshine is strongest and where the sun's ultraviolet radiation is most effective in transforming O_2 into O_3, the two-atom (diatomic) molecules of oxygen into the three-atom (triatomic) variety. In this process some of the two-atom kind are first made monatomic, before these single atoms are attached to ordinary oxygen, thereby forming ozone.

Confusingly this ozone then breaks down quite readily, and speedily, to be reformed as conventional oxygen, and the process of breakdown and build-up is continuous within the stratosphere. Not only is ozone formed by UV radiation acting on ordinary oxygen, but ozone is also destroyed by UV radiation of a slightly longer wavelength. The balance has looked after itself, as it were, until mankind came along to upset it – about which more in a moment. The natural and ongoing procedure absorbs much, or most, of the sun's UV radiation which, if it reached the Earth, would be extremely damaging, notably by acting as a carcinogen on human skin and in the creation of cataracts. UV forms only a minute proportion of transmitted solar energy, but it has great potential for harm. (Strictly speaking, it is the ultraviolet radiation with wavelengths less than about 325 nanometres which is so unwelcome.)

The harm to humans has received considerable publicity. Skin cancers, on the increase in many areas, most famously in Australia, have alerted people to the dangerous effect of the sun. It would seem that almost everything we like doing, such as toasting ourselves a different colour or merely enjoying the glow of sunshine, can be damaging. We therefore use sunblockers – by the many thousands of tons annually – partly because the natural blocker of an ozone layer is inadequate and becoming more so. If our faces, in particular, are much exposed to sunshine, and also the backs of our necks, this human skin may start to resemble a dried-up prune as extra testament to the darker side of solar radiation. Various eye diseases, as well as cataracts, are known to be linked to a sun which undoubtedly gives us life but makes us pay a price for its beneficence. The form of ultraviolet radiation known as UV-B is the kind which causes sunburn and malignant melanoma. The peak levels in the regions where we live have certainly been rising recently, with New Zealand, for example, registering a 12% increase between 1990 and the end of that decade.

Increased UV radiation has other effects than those on humans, such as hindering the growth of marine plankton. We should never forget that the pelagic realm, the largest ecosystem of them all, lies at the base of much of life on Earth. Harm done to plankton may be much more damaging to humans, however indirectly, than the direct, albeit carcinogenic, effects of UV radiation. There is also talk that depletion of the ozone shield may damage DNA and therefore cause genetic mutations, for instance in maize and other plants.

The low-level atmospheric ozone is chemically identical to the stratospheric kind, and will also screen out ultraviolet radiation, but it is an unwelcome offspring of pollution, notably from cars and trucks. Sunlight acts upon their exhaust gases to create what is called photochemical smog, a hotch-potch of different substances, one of which is ozone and all of which make an unpleasant contribution to modern urban life. The US Clean Air Act of 1990 required oxygen-rich compounds to be added to fuel in the hope of reducing carbon monoxide production which, otherwise, could lead to ozone formation.

Peak city levels of ozone in the United States did fall by 10% between

1986 and 1997, but a variety of other amendments, including cleaner-burning fuel, have been given the credit for this drop, a reduction which occurred despite an increase in vehicle numbers during those 11 years. The principal 'oxygenator' added to fuel was MTBE (helpful abbreviation for methyl tertiary-butyl ether), but its incorporation was quickly followed by complaints of headaches and nausea, presumably over and above those created by the unwelcome smog which MTBE was attempting to reduce. A definitive study by University of California researchers found that 'MTBE and other oxygenates [had no] significant effect on exhaust emissions from advanced technology vehicles'. In other words it was preferable to make better cars and better fuels than additives with unwelcome side effects.

The transformation of ordinary oxygen into ozone always needs energy. Whether supplied by UV radiation, or radioactive bombardment, or electrical discharge, the resultant ozone is the same. As for translocation of stratospheric ozone (formed by ultraviolet) and tropospheric ozone (created by human industry) there is a little movement between the two zones, but this is poorly known. Some civil airliners, flying at their normal cruising altitudes between 6.2 and 7.5 miles above the Earth, have recently been used to collect valuable data about possible mixing between these atmospheric regions.

Oxygen, the most abundant element on Earth – even if only one-fifth of the atmosphere – is crucial for most life forms. It is formed from water during the process of photosynthesis, and most of it is recycled back to water during the quite different process of respiration. Estimates of this perpetual recycling, which will continue so long as there is life, show that there is a complete turnover of this crucial gas every 1,200 years. Living forms make it and living forms use it. However, oxygen's properties become quite different when an extra atom has been added to each of its molecules.

The subsequent gas is still colourless, but has a pungent smell, unlike odourless oxygen, and it can be harmful. Commercially it is used as an ingredient of bleaches and sterilisers, and even for removing other smells as it can oxidise unwelcome compounds. Seaside resorts, notably in the 19th century, used to proclaim the pungency of their ozone-rich environments but, in the wake of so much negative publicity about the gas, they have tended to drop this form of commendation. In any case the ozone quantities were too small to be detectable by the human nose. Victorian seaside addicts and enthusiasts were probably, and most contentedly, sniffing the iodine released by rotting seaweed.

Perhaps ozone should be equated with knives, or glass, or morphine, or any of a host of items which are either useful and beneficial or can be severely damaging, and possibly lethal, in other circumstances. (Even pure oxygen can be lethal to humans, as scuba divers learned to their cost before realising that it must be diluted with some other gas, much as nitrogen dilutes it within ordinary air.) Up in the stratosphere ozone serves as covering, protecting us from unwelcome radiation and preventing heat loss from our planet. Down where we live, in the lower troposphere, the gas is still able to absorb ultraviolet radiation but, as powerfully indicated by its use as bleaching agent, it is inimical to life. Ozone is therefore good, higher up, and bad, lower down. The major trouble, and why the ozone word hits headlines quite so often, is that human activity is not only diminishing the quantity higher up but adding to the quantity lower down. On both counts we are therefore doing wrong.

During the mid-1980s scientists of the British Antarctic Survey working from their southern bases realised that stratospheric ozone above their general area in springtime was diminishing, and about half of the usual quantity had already disappeared. Such measurements had only begun in the 1950s when the ozone layer did seem to be stable, and it remained comfortingly so for three decades. The subsequent realisation that things were altering was not immediately convincing to all others, mainly because ozone levels can fluctuate by 30% from day to day, or even from one year to another by 10% or so.

The missing ozone was tellingly described in 1985 as a hole, and this alarming revelation, of an absent portion of the stratosphere, was instantly attributed to human activity. Much earlier it had been predicted that such a depletion might happen, perhaps being caused by high-flying aircraft (like the very highest, both civil and military), or the increased use of artificial fertilisers, or by chlorofluorocarbons (CFCs), these being the product of various industrial processes, or by a combination of them all. Aircraft and fertilisers were known to be potential sources of oxides of nitrogen, which also act as powerful ozone destroyers. CFCs were known to interfere with ozone creation by forming destructive chlorine compounds which react readily with ozone, each chlorine atom being able to destroy many molecules of ozone.

The protective and triatomic form of oxygen was therefore seen to be in jeopardy. The earlier warning had come from a critical paper, published in 1974 by two University of California scientists. It was the first to draw concerned attention to this undoubted problem, and to its possibilities. These are made particularly severe by the fact that some man-made compounds related to the CFCs have an atmospheric lifetime of more than 1,000 years.

Amazingly, in a world accustomed to bureaucratic aeons before global strategies are ever implemented, an international meeting at Montreal in 1987 – only two years after the ozone hole discovery – determinedly agreed that production of CFCs should be immediately restricted. Quite apart from their role as ozone destroyers a single CFC molecule within the atmosphere has a greenhouse effect 5,000–10,000 times greater than each added molecule of CO_2. It was therefore decided that, as from 1 January 1989, CFC consumption should be frozen at 1986 levels and this quantity should be reduced by 50% within ten years.

At Montreal it was even recommended by all parties that a total ban on these undoubtedly useful but intensely damaging chemicals should be the final aim. Such prompt and welcoming unanimity has been called the world's most successful piece of international environmental legislation (and there is much more about it in *Ozone Diplomacy* by Richard Benedick, the 'definitive story' of this achievement). Within five years of that 'Montreal Protocol on Substances that Deplete the Ozone Layer' the global production of the most harmful CFCs had fallen by 40%.

Unfortunately not all of the 163 signatories to the Montreal agreement, and to its subsequent amendments ratified in London and Copenhagen in 1990 and 1992, have been as good as their word, with Russia as a major villain. Its seven major CFC factories are not only producing for the internal Russian market, largely for refrigerators and aerosol cans, a procedure phased out in the West by the late 1980s; but also via illegal smuggling to other markets, notably those in some of the former Soviet states. (As a happy 'Exam howler' put it, collected by a teacher and then published in *The Biologist* under that heading: 'Aerseholes release CFCs'.) Even in Florida it has been estimated that 10,000 tons of these chemicals were illegally imported between 1994 and 1996, the street value of this substance allegedly not much lower than cocaine. Black market demand is high because owners of old cars prefer the cheaper CFC-based coolants for their air conditioners rather than the legal and higher-priced alternatives.

Before the break-up of the Soviet Union one-third of world CFC production originated in Russia, and that global proportion has since risen. Its factories should have closed early in 1996 but, after failing to do so and with Russia blaming the poor state of its economy, wealthier nations were asked by the World Bank to provide helpful funding. Some contributed, but insufficiently. By the end of 1997 over 300,000 tons of CFCs were still being manufactured, with the producers listed as: Russia

47%, China 28%, India 7%, Korea 7%, Venezuela 5%, Brazil 3% and Mexico 3%. What is being called environmental crime is very much in business.

Another most potent ozone destroyer is methyl bromide, a pesticide and herbicide much deployed around the world, notably against nematodes, fungi and weeds. Soils are fumigated with it before certain valuable crops are planted, such as strawberries, tomatoes, grapes and flowers. Californian strawberries alone have been using 3% of the world's supply. When this gas escapes into the atmosphere it releases bromine compounds which destroy ozone. Molecule for molecule it is even more destructive than CFCs and is thought to be responsible for 10% of the ozone layer's destruction. This uncertainty arises from curiosity why such a massive use of methyl bromide, coupled with its destructive power, is not more damaging to the ozone layer. In any case the wealthier nations have agreed to phase it out by 2005, while developing nations have until 2015 to follow suit. Israel is currently the world's largest manufacturer, but China is about to usurp that position. Unfortunately various prosperous areas, such as Italy and Spain, each intensely aware of the pesticide's virtues with their precious crops, have been slow to accept the strictures, requesting that these be watered down. Fortunately there are alternatives to methyl bromide, allegedly cheaper, such as covering the soil in plastic, watering the earth beneath this covering, and letting sunshine boil the water to sterilise the soil.

With so much attention focussed on CFCs and on other ozone depletors such as methyl bromide, one particular renegade has managed to escape much of the attention. Emissions of halon-1211, a major ozone destroyer (and second in line after CFCs), have been quietly rising by some 25% since 1987. This chemical is used in fire extinguishers, and China produces 90% of it. Unfortunately the emissions of this halon are even greater – by 50% – than the documented evidence of its production. The ozone layer's recovery is therefore being set back 'several years', according to investigators, owing to this single error. Removing CFCs and methyl bromide without attending to halon is like focusing on wolves without concern that a tiger (or dragon) is also on the prowl.

There is also halon-1301, widely used in fire extinguishers – notably for aircraft. The commercial airlines currently carry 700 tons of this swift-acting, non-toxic chemical, and actually need it to acquire an airworthiness certificate. There is thought to be no safe alternative, but the manufacture of this halon is ceasing in 2006 as part of a policy to protect the ozone layer. Plainly we want aircraft fires to be quickly and safely extinguished. Equally plainly we want no further reduction of the ozone layer. No less blatantly these two conflicting compulsions will have to be addressed so that a solution, as yet invisible, will be found.

One way and another, what with outright disagreement, lack of cash for compensation, lobbying from manufacturers, lack of suitable pesticide alternatives, clever smuggling of banned chemicals, and determined miscreants countering progress, there was not the jubilation there might have been when delegates reassembled in 1998 for the tenth anniversary of the Montreal Protocol. The original international agreement had been 'a leap of faith', according to the chief US negotiator, a man who later admitted that reductions in ozone destroyers 'were mandated with the full knowledge that alternatives were not available'.

The Montreal Protocol's 1999 meeting took place in Beijing, China, at the end of the year. It was then revealed by representatives from UNEP, the UN Environment Programme, that worldwide production of ozone-damaging chemicals, such as CFCs and halons, had been rising, with CFCs up by 5% since 1997 and halons by 10%. China, the conference's host country, was labelled as principal villain. During the 13 years since the original Montreal conference, when so many nations agreed so speedily to phase out production of ozone-destroyers, China has actually increased its own production of them fourfold. Western nations have phased out their manufacture of these unwelcome products and were asked at the conference to provide up to $500 million to help China and its customers arrange for a lessening of both manufacture and supply. A total of one billion dollars has already been provided by Western nations to help developing countries make and use alternative, less damaging products for their refrigerators, aerosols and fire extinguishers. The conference's principal Chinese delegate asserted that manufacture and use of the harmful products would have totally disappeared in his country by 2010.

Causing pessimism in late 1998 the Antarctic ozone hole had been reported to be bigger than ever. It was also forming much earlier in the year, starting in June as a ring around the continent rather than a hole, this ring then working inwards to form the hole. The tenth anniversary jubilations were indeed somewhat muted. The area of missing ozone, the fundamental cause of Montreal's deliberations, had grown to become a gap three times the size of the United States. Worse still, and complementing the bad Antarctic news first reported in the mid-1980s, the Arctic's equivalent layer was also reported during the mid-1990s to be thinning. Record low temperatures in the northern stratosphere have caused researchers to fear that ozone destruction may, in consequence, be exacerbated.

Part of the trouble is that the cold air vortex which forms over Antarctica every winter (and similarly over the Arctic six months later) had fallen in temperature below -80 degrees C. in the north and down to -86 degrees in the south, some four degrees colder than had been customary. It seems contrary that climate near the ground is growing warmer, which it is, while the lower stratosphere is cooling, but these facts, like so many others in the climate business, are not fully understood. It is also not understood why colder temperatures should worsen the depletion of ozone, but that is happening. Perhaps the increasing cold causes stratospheric clouds to last longer, and perhaps such clouds remove the oxides of nitrogen which normally deactivate the ozone-destroying chlorine molecules, or perhaps there is a mixture of much else.

Nevertheless there are two certainties: human beings are attempting to be less damaging to the ozone layer and, despite these efforts, the ozone hole is still increasing. Had Montreal not happened it is generally appreciated that the hole would be bigger still. It is also believed that a different kind of human activity may also, along with CFCs, etc., be to blame for the ozone's disappearance, namely the production of greenhouse gases. These trap heat near the ground and stop it rising to warm the stratosphere. The upper atmospheric layer therefore cools and forms more clouds. The clouds then permit greater survival of chlorine and therefore greater destruction of ozone. In short it is humans, yet again, who are causing the alterations and must mend their ways.

Everyone applauded Montreal, even those who could not, or would not, obey its commendations. It was a speedy and global response to an environmental issue, the most impressive there has ever been. As for all the greenhouse gases, about which much more in a moment, their possible reduction is a great deal more complex, but it is being addressed, even if not universally. The fact that the gases damage Earth's protective ozone layer is one more reason why they need to be curtailed.

Montreal was such a buoyant event, with the planet's governmental representatives acting in such concert about the planet's affairs, that immediate reward was anticipated. The ozone's thinning would, it was felt, be stopped in its tracks, with the layer back to normal about as speedily as the Montreal assembly had been arranged. Sadly this has not proved to be the case. A depressing forecast, published in June 1998 by the World Meteorological Organisation, stated that signs of the ozone layer's recovery may 'not become apparent for the next two decades' and its full recovery is 'not expected to occur' until the middle of the 21st century. Despite such discouragement the 1987 protocol to phase out ozone-depleting substances 'is clearly working', according to WMO's secretary-general when speaking in 1998. It just is not working fast enough, or with sufficient vigour.

Carbon dioxide There is nothing new about the so-called greenhouse effect warming Earth. If it did not already exist naturally we would be in a parlous state, with global temperatures about 33 degrees C. cooler than currently exist. For most of the time, and in most places, the terrestrial temperature would be 19 degrees below freezing point instead of the more agreeable average of 15 degrees above it. For all the negative talk these days about the threat exercised by greenhouse warming it should be remembered that this phenomenon is of extreme benefit. The trouble, unlike Mae West's dictum in another context, is that it is possible to have too much of a good thing.

There is also nothing new about human understanding of greenhouse warming and what it means for planet Earth, the phenomenon having been outlined by the French mathematician Jean Baptiste Fourier in 1827, and then more accurately described, first by John Tyndall, a Briton, and particularly by Svante Arrhenius, of Sweden, later in the 19th century.

There is also nothing new about carbon dioxide being a prime component of global warming. This gas has always existed in Earth's atmosphere and has been crucial in making the planet agreeably warm for life. It must also have been a frequent cause of earlier climatic change, whenever carbon embedded in rocks became once again exposed or, conversely, when much carbon was being extracted from the atmosphere and sealed in sediment, as with the carbonates of chalk. Professor Arrhenius was writing as much in the *Revue Générale des Sciences* during the 1890s: 'If the amount of carbonic acid [in the atmosphere] were diminished by a little more than half, the temperature would be lowered by about 4–5 degrees C. while an increase to two or three times the present amount would raise the temperature about 5–8 degrees, corresponding to the conditions of Glacial and Eocene times respectively.' There was then no mention, from him or anyone else, that human beings might themselves be adding to that carbonic acid and perhaps nudging the planet's warmth in the direction of another Eocene.

A word first about carbon, the linchpin of life on Earth. It is present in all organic substances and is one of Earth's commonest elements. It would be possible for life to be based on other elements, this supposition perhaps to be proven when life forms from elsewhere are encountered, but carbon is particularly suitable, mainly because of its valency of four. Each carbon atom has four bonds. It can therefore easily link with other atoms to create the huge molecules which are so characteristic of carbohydrates, these being essential components of all living organisms. Sugar, cellulose, starch and glycogen are some of the simplest and most fundamental to be found wherever life is found. Carbon by itself can exist (allotropically) in more than one form, such as diamond and graphite, and (amorphously) as lampblack, but carbon in organic material is part of an uncounted, and probably uncountable, number of different molecules, each fulfilling their different roles.

What is now new about carbon dioxide, and aggressively so, is the current awareness that human activity has definitely caused it to exist at far higher levels in the atmosphere, along with other greenhouse gases. This understanding first surfaced properly in the 1940s when the initial calculations were made linking the burning of fossil fuels with increased CO_2. It was then realised that the entirely natural and extra warmth of 33 degrees, first comprehended in the 19th century, might well be raised by one or two degrees as a result of human industry, largely from the burning of fossil fuels in the 20th century.

Such an increment can seem small, and is small relative to that natural greenhouse gain, but climate is a fickle matter, with minor alterations having major consequences, like a final snowflake triggering an avalanche. It is now well understood, and is rightly of concern, that much of what we do, whether burning oil and coal, letting natural gas escape, clearing and burning forests, keeping ruminants, planting various crops, using fertilisers, disposing of countless unwanted items, manufacturing cement, creating CFCs, and even breathing can add to the greenhousing of our planet, and therefore to that little but mightily important rise in temperature.

For purists the term greenhouse effect is incorrect, there being a difference between the container of a greenhouse and the gradual thinning of an atmosphere, with air cooling according to its height above the surface, but the name serves well enough and is unlikely to go away. Almost anything can serve in greenhouse fashion. A tent can become unbearably hot when the sun is shining on it, and so can a car, killing off dogs and babies inside unless sufficient ventilation is arranged.

Essentially the sun's short-wave radiation, at wavelengths from 0.3 to 2.0 micrometres, is unimpaired (save for about 6% scattered back to space) as it travels through the atmosphere to reach and warm Earth's surface. About 10% of that received heat is then radiated back from Earth into the atmosphere, but in its rejection it has been transformed, its wavelengths having been increased to 10–15 micrometres. There is nothing magical about this transformation. The sun is hot and therefore produces short-wave radiation. The Earth, however much warmed by the sun, is relatively cool and therefore produces long-wave radiation. As is well known, ordinary domestic electric fires become brighter and change colour as they warm up, with altering wavelengths from that increased heat visibly conspicuous. The longer form of wavelength emitted from a warmed up Earth cannot pass readily through air, and is largely absorbed by it, thus heating up the atmosphere.

Some of this acquired heat is then re-radiated back to Earth and some of it is lost to space, but the net effect is to keep Earth much warmer than it would be without this all-encircling covering of greenhouse gas. As for the heat which is returned again to the Earth's surface a proportion of it is then radiated once more, and once again as long-wave energy to be absorbed by the atmosphere. The net result of all the original heating, minus the amount radiated upwards but plus the amount radiated back again, is that each square yard of Earth's surface is warmed by about 200 watts of energy. That is a comprehensible quantity, as all of us know about 60 or 100 watt light bulbs or even single-bar electric fires of 1,000 watts. The quantity of heat received by Earth only becomes incomprehensible when multiplied by the number of square yards on Earth's surface, with pi multiplied by the diameter-squared being the surface area of a sphere.

Carbon dioxide, always created by the combustion of coal, oil and natural gas, has received most publicity, but is only one of some 20 gases responsible for the greenhouse effect. Others making major contributions are methane, nitrous oxide, the CFCs already mentioned, and tropospheric ozone (as against the stratospheric ozone much higher above the Earth). However significant their effects these gases are minor in their proportion of the atmosphere, adding up to less than 1% of it. Water vapour also serves as a greenhouse gas, and is the most important, with air either bone-dry of it or fully saturated, but its atmospheric quantity has received less attention because it has not been directly altered by human activity.

The bulk – 57% – of greenhouse gas emissions created by human beings, according to a summary by O. Green and J. Salt in *Ecodecision*, comes from the burning of fossil fuels and from industry in general. The destruction of forests, coupled with their burning, produces a further 8%. Agriculture, including rice growing, the keeping of ruminants and the use of fertilisers, provides 9%, whereas waste management, including disposal and incineration, produces a further 5%. The manufacture of cement creates 1%, of CFCs 11.5%, and a miscellany of other sources the final 8.5%. No wonder, therefore, that CFCs were jumped on, their proportion so considerable and their production so unnecessary relative to rice, cattle, cement, fertilisers, and so forth. Fossil fuel consumption does indeed create the largest share, with cars in particular having an iron grip upon the modern human psyche. Suggest reducing their personal transportation vehicles to any group of Americans, as was said at a greenhouse conference, and 'they all immediately require cochlear implants'. They also tend to lose their hearing if a reduction in air travel is suggested, but this is a lesser issue as aircraft worldwide only produce 3% of the carbon dioxide created by burning hydrocarbons.

The United States is the world's leader in greenhouse gas emissions, producing 20% of the total from 4% of the world's population. The next 11 highest producers are, according to some recent figures and in descending order, China, Russia, Brazil, India, Japan, Germany, United Kingdom, Indonesia, France, Italy and Canada. The British contribution is 2.7% of the total from 0.9% of the world's people. Therefore Britons are about 60% as gas-productive per individual as are Americans. A quite different ordering occurs if the greenhouse gas production per head of population is listed. Canada then displaces the United States as top producer, being followed by Ivory Coast, Brazil, USA, Australia, Saudi Arabia, Netherlands, Germany, UK, Russia and Colombia. In this kind of listing, and for the two most populous countries, China comes 24th and India 25th.

During each year the Canadians are producing 4.5 tons per person, the British 2.7, the Chinese 0.34, and the Indians 0.28 tons. With the poorer nations wishing to emulate the wealthier, and with the richer nations unwilling to forsake their current prosperity, it would seem inevitable that gas emissions are set to rise. Any percentage increase per person in either China or India, for example, has to be multiplied by over a billion in each case to assess the additional contribution to global gas. Even on today's figures, although Canada is lavishly emitting 112 million tons, India is – far less lavishly per individual – producing 280 million and China 384 million tons. All such colossal figures are likely to become yet more daunting, not only as the rich get richer but as the less rich are able to improve their lot. Currently, as an easy guide to CO_2 emission, one-sixth of the world produces 55% of it while the one-sixth at the other end of the economic scale produces 3%. As for the remaining four-sixths of the world's population, neither grossly rich nor grossly poor, they produce the remaining 42%.

China merits a paragraph (at the very least) to itself. It suffers from a huge demand for coal, and mines one third of the world's total quantity. Its smogs and general pollution are visibly awful. Lung diseases are the country's major cause of death. Not only is it the most populous nation, but that population is still increasing and in partnership with massive economic growth (above 10% most years). The place has been called 'a one-nation destroyer of global climate'. Simultaneously, and on the flip side, China is desperately attempting to put its house in better order. A January 2000 'Focus' review in *New Scientist* stated that it closed down 31,000 coal mines in 1999, has shut down 60,000 inefficient industrial boilers, and has cut coal use by 30% to about one billion tons. Overall the Chinese economy has doubled in the past 15 years while air pollution has only risen by 50%. Crop yields have been reduced by 30% in much of the country due to acid rain and haze. Plainly there is still much to do, but China has, at last, set about the task. The world's CO_2 problem is not a matter for China alone, but this one nation is responsible for a considerable disproportion of it.

The gases principally involved in the greenhouse cause of global warming – carbon dioxide, methane, nitrous oxide, ozone, CFCs – all absorb the long-wave radiation from Earth, but their proportions in the atmosphere have each altered in recent years. David D. Kemp, in *Global Environmental Issues*, has listed the average figures for the 100 years from 1880 to 1980. Initially these were CO_2 66%, CH_4 15%, CFCs 6%, NO_2 3% and all others, including ozone, 10%. By the 1980s these ratios had altered to CO_2 49%, CH_4 18%, CFCs 14%, NO_2 6% and others 13%. The reduced percentage of carbon dioxide can make it seem as if its production has been reduced, but only its share of the greenhouse gases, and certainly not its creation, has been lessened.

As a further aggravating point, also increasing the quantity of CO_2, much forest has been destroyed, notably during the 20th century's latter half. According to Norman Myers, arch-enthusiast on behalf of trees, forests covered 20 million square miles in 1950, a figure which has since shrunk to 13 million. 'At the current rate of deforestation,' he wrote in a July 1999 book review, we are 'likely to witness the demise of most tropical forests within the lifetimes of most readers of this journal.' The amount of carbon dioxide being consumed by plants during photosynthesis is therefore being reduced, causing more of it to remain in the atmosphere, whether or not more or less of it is being produced by the human consumption of fossil fuels. Methane and the CFCs have both risen considerably in the years since 1980, thus squeezing the proportion of CO_2

but not, to make this point again, its actual quantity.

All such additions can make us think these extras are somehow reducing the air's two main constituents, namely oxygen and nitrogen. If only one fact lingers from schooldays about Earth's atmosphere it is that oxygen and nitrogen exist as one-fifth and four-fifths of its volume respectively. So is this basic piece of remembered information no longer true? If still valid how do all the extra greenhouse gases fit into this compact arrangement of four-fifths plus one-fifth equalling five-fifths? The answer is that, for all their power of altering Earth's temperature, the extra gases occur in minute quantities. In percentage terms oxygen and nitrogen are generally written as 78% and 21%, leaving a casual 1% for all the extras. In truth this single percentage point is excessive as the other gases exist most modestly, so much so that they are usually written in parts per million. Their diminutive contributions make most sense if *all* the gases of the atmosphere are written in parts per million, including the two most powerfully represented.

Nitrogen	780,000
Oxygen	210,000
Water Vapour	20,000 (maximum, but can be 0)
Ozone	1,000 (maximum, but can be 0)
Carbon Dioxide	360
Methane	1.8
Nitrous Oxide	0.3
CFCs	0.001

This list totals more than 100 per cent if the maximum figures are assumed, but the figures are there to show scale. Plainly, if there is much moisture or ozone, the relative quantities of oxygen and nitrogen have to be very slightly reduced. If, on the other hand, there is very little ozone or water in the atmosphere, there will have to be slightly more of the other gases. The fact that CFCs are one part per *billion* can make us wonder about all the fuss, until remembering their extreme ability to damage all-important ozone and their considerable power as greenhouse gases. Methane, although in lesser quantities than carbon dioxide, is about 20 times more effective as a greenhouse gas per molecule than is CO_2. The tremendous proportions of oxygen and nitrogen are irrelevant to our planet's warming because they neither emit nor absorb terrestrial radiation.

Quite the most worrying fact about carbon dioxide is its recent remorseless rise, a certain consequence of human activity. In pre-industrial times about 280 parts per million of the atmosphere were CO_2. Excellent confirmation of carbon dioxide's atmospheric rise has come from Antarctica, this place serving as a storage refrigerator for ancient history. One surprise for Antarctic visitors is the steady pop-gun crackling from melting ice when its trapped, and highly pressurised, air bubbles are released from their confinement, this imprisonment having lasted for centuries or millennia. In March 1999 a paper in *Nature*, written by a dozen Swiss and American scientists, summarised work on these air bubbles which had started in 1958. Pre-industrialisation atmospheric CO_2 had been 280 parts per million, but this quantity increased from about 200 to 270 during the period between the Last Glacial Maximum (20,000 years ago) and the beginning of the Holocene (the most recent and post-glacial epoch, beginning 11,000 years ago).

In 1958, when the trapped air work began, the quantity of CO_2 was 315 parts per million. Since 1958 the amount has been steadily rising, and it reached 364 by 1997. Worse still, for optimism, no one is expecting its level to reach some form of plateau in the foreseeable future. Demographers are estimating that today's global human population of 6 billion will, probably, not go much higher than 11 billion, and will reach that figure in the middle of the 21st century; but no one has been affirming that the wish of humans to want more, to consume more, and therefore, almost certainly, to create more CO_2 while fulfilling these demands, will somehow be reduced.

If there is one general sentiment among all of Earth's human inhabitants it is that today's poorer citizens, and certainly the poorest, should achieve a higher standard of living; but, if the 11 billion people of a stable population as from the 2050s all succeed in living as Americans are living today, that egalitarian world will be producing 46 times as much greenhouse gas as the unequal world is currently achieving. This thought is daunting, quite apart from the accompanying presumption that Americans and their like may then be consuming even more – of every-thing? – than they are today.

Carbon dioxide is not the only greenhouse gas to have increased recently. Methane, much produced from the bowels of a swollen number of ruminants, has multiplied about 2.5 times from its pre-industrial age quantity of approximately 0.7 parts per million. Ice-core work in Greenland has indicated that this original quantity did not vary for 2,000 years – until industry began. Nitrous oxide, another product, as with CO_2, of burning hydrocarbons, has also increased, but more modestly. As for CFCs their quantity has risen from zero, as it was in the old days, to their current level. Not one of these several greenhouse gases immediately disappears once it has reached the atmosphere and has started upon its warming role. Methane's lifetime is 10–12 years, but ordinary CFCs can last for much longer, and CO_2 plus NO_2 each survive in excess of a century. The problems these gases cause in global warming will not therefore vanish speedily, even if their production could magically and instantly be stopped right now.

Kyoto Stopping is an impossibility. Some form of curtailment is not only feasible but might, and should, occur. At a meeting in Kyoto, Japan, during December 1997 the United Nations Framework Convention on Climate Change chose targets for reductions in greenhouse gas emissions. In essence the Annex 1 countries of 38 industrialised communities agreed, in principle, to a 5.2% reduction by 2010 relative to 1990. These well-developed nations cause 57% of such emissions at present. Unfortunately, although they are the major offenders and intend to make reductions, it is expected that most future growth in greenhouse gas production will occur in south-east Asia and Latin America. These fast-developing regions were not signatories to the Kyoto convention.

Those who did sign in Japan, such as various members of the European Union, were subsequently more specific about their obligations. In June 1998 Germany announced a 22.5% reduction from 1990 levels before 2012, Denmark and Austria a 21% cut by 2010, the UK a 12% cut by 2012, and France and Finland no cut at all but a stabilisation at 1990 levels. Contrarily, and for less developed members of the EU, it was decided that Portugal should be allowed to raise its 1990 emissions by 24%, Greece by 23%, Luxembourg by 30%, and Spain by 15%. These increases, coupled with the reductions, add up to a reduction of 8% from 1990 levels before the year 2012. This, in brief, was the EU commitment made at Kyoto during the famous meeting of 1997.

Promises are easier to make than keep. Even at Kyoto, and certainly afterwards, certain countries, such as the US and Japan, wished to amend the general ruling. They preferred that the reductions should be based on 1995 emissions rather than those of 1990. As both these countries, particularly Japan, were producing more greenhouse gases in that later year it was plainly easier to make reductions from a higher figure than a lower one.

There was also disagreement, notably in the US, about the arguments underpinning Kyoto. Increases in carbon dioxide levels actually 'benefit' Earth, rather than harm it, according to a petition signed in April 1998 by

15,000 science graduates, including 6,000 PhDs. The chairman of the US House of Representatives Science Committee echoed this point: 'I don't think we should risk harming the American economy, and by extension the American people, by ratifying the Kyoto treaty.' The Japanese government argued that the EU proposal for a 15% cut from 1990 levels was 'unrealistic'. Australia, one of the world's highest atmospheric polluters per head of population, with 90% of its electricity coal-generated, argued against any form of blanket ban, preferring that each country sets its own targets.

There was a prompt backlash to the signature collection, with others affirming that 15,000 formed 'a small proportion' of those who were mailed, that the US possesses 'more than half a million' science or engineering PhDs, and 'ten million' individuals with first degrees in those two subjects. Arthur Robinson, president of the privately funded Oregon Institute of Science and Medicine, would not say how many requests for signature had been despatched but did state, in a paper he co-authored, that the release of more carbon dioxide 'will help to maintain and improve the health, longevity, prosperity and productivity of all people'.

There has since been backlash to the backlash. The Union of Concerned Scientists considers the US could reduce greenhouse gas emissions by nearly twice the 7% called for in the Kyoto protocol. That figure could become 13% below 1990 levels by 'using domestic actions that will have zero or negative net costs'. The nation's energy bill 'could be cut by $50–90 billion a year' by:

(1) switching to natural gas instead of electricity for residential use;
(2) replacing coal-burning plants by others using natural gas;
(3) relying increasingly on fuels such as ethanol for transport.

The US Department of Energy has itself argued that carbon emissions could be cut to 1990 levels with no 'net cost' to the nation's economy. Critics of this optimism have replied that such a reduction would charge $2,000 per American household per year in 2010. The American Petroleum Institute, somewhat inevitably weighing in on the side of hydrocarbons, has estimated that capping carbon emissions could reduce the gross domestic product by 2–4%, or $200–350 billion a year. Global Climate Coalition, the principal industry group within the US lobbying against Kyoto's strictures, initially included the biggest car manufacturers. Then, in 1999, Ford withdrew from GCC and in January 2000 DaimlerChrysler followed suit, the recently merged company of Daimler

Benz and Chrysler. General Motors is now the only major car maker still supporting GCC's position on global warming.

Eleven months after Kyoto the climate negotiators met again, this time in Buenos Aires, and 150 nations were represented. One year earlier, as *New Scientist* phrased it, the oil industry 'had a face of flint', but this unwelcome visage had started to change. BP, the oil giant, had announced plans to cut its own emissions by 10% before 2010. Shell then followed suit, and was pleased to report it would spend $30 million on solar panels for rural South African homes, the world's largest project of its kind.

Kyoto had let developing countries off the hook, and the US argued in Argentina that they should have fixed emission limits before the US would be prepared to cut its own production of greenhouse gas. A compromise by the World Resources Institute, of Washington, DC, suggested that these developing nations should reduce the amount of carbon produced per dollar of their gross national product rather than abide by some ceiling. By this criterion some were already doing better than others. Whereas India's carbon production per dollar had increased by 29% since 1980, and Malaysia's had grown by 58%, that of China had dropped by 47% during the same period. Aubrey Meyer, of London's Global Commons Institute, said: 'If we do not set absolute limits on emissions, we are ignoring the physics of global warming.'

As for those physics they are certainly not straightforward. The billions of tons of carbon dioxide which humans generate each year does not all turn up in the atmosphere. Where the rest goes is a mystery, with scientists referring to a sink which is absorbing/receiving the missing CO_2. It is believed by some, such as the authors of a paper in *Science* in October 1998, that trees, by growing faster and larger in the presence of more CO_2, are mopping up much of this extra gas, perhaps 40% of the quantity which has gone astray. It is even thought, by a team known as the Climate Modeling Consortium, that the US forests may be soaking up 1.5 billion tons of carbon dioxide a year, which is more or less what the US is producing. Less than a year later a team at the Woods Hole Research Center, Massachusetts, came up with a different estimate, this stating – also in *Science* – that US ecosystems were 'stashing' no more than one-fifth of the CMC figure. 'Forests are not going to save us,' concluded an investigator at Washington's Carnegie Institution who had examined both sets of estimates.

Europe's forests, less receptively, are thought to be absorbing between 120 and 280 million tons of carbon more than they are releasing. This quantity is either a sixth or a third of the 800 million tons of CO_2 currently produced on the continent. Britain does much less well than the European average, mainly for being one of the world's least forested countries and one of the most crowded, certainly with trucks and cars. Its trees only absorb about 2% of the nation's CO_2 emissions. In any case not everyone is sanguine about forest absorption rates. Other investigators point more favourably towards the oceans as possible explanation for the missing sink. There is as much dissolved organic material in the oceans as there is CO_2 in the atmosphere, but this oceanic role, as has become such a refrain in this book, is poorly understood. Approximately half of all plant production occurs in the oceans, a fact easily forgotten or mis-remembered by terrestrial individuals such as humans.

What is known for sure is that oceans transport massive amounts of heat around the globe. Western Europeans, notably the English, were amazed when they first travelled to similar latitudes in the New World and discovered how cold these were in comparison with home. Boston and Cape Cod, initial home of many emigrants, lie on the same parallel as northern Portugal, and yet are perishing in winter. (The Pilgrim Fathers finally got everyone ashore on 26 December, and half the *Mayflower*'s clientèle died during the following three months, with cold a powerful factor in their demise.) New York and Chesapeake Bay are even further south, and can freeze in a manner which southern England can never emulate. As for the coast of Labrador, more or less equivalent to the British Isles, this portion of Canada might be situated on a different planet for all its similarity.

Gradually it was learned that Britain and coastal Europe, rather than the new world of the western hemisphere, were exceptional, with the Gulf Stream being responsible, this flow bringing extra warmth now calculated as ten billion megawatts. Its heat warms the winds which blow so benignly, even if blusteringly and damply, across Europe. As for the beneficent ocean current, it then cools to travel south about 6,000 feet below the surface. This massive circulation, seemingly unalterable and permanent, in fact relies upon very small density differences in the water. It could be – this dread thought first indicated by Henry Stommel in 1961 – that a small amount of additional fresh water, generated by increased rainfall, might tip the balance in favour of a gross amendment to the circulation. Western Europe would then know winters much as Cape Cod and Massachusetts have always known. By 1996 some climate models in computers were also suggesting that rising water temperatures, caused by global warming, could reduce ocean circulation, and would simultaneously cut the sea's ability to store CO_2.

In March 1999 David W. Schindler, of the University of Alberta, summed up in *Nature* much of the current thinking on the world's carbon cycling, and where much of the missing carbon might be hiding. This element is undoubtedly being produced prolifically, mainly in northern latitudes. With half this carbon staying in the atmosphere, and a quarter in the oceans, it is the remaining quarter which makes for mystery. This sounds a lot, but it is not easy to find, and Schindler has addressed its possible hiding places.

First, the oceans, with CO_2 perhaps being transported from the north via deep water to the south where it is vented into the atmosphere. Second, northern trees have been growing faster than expected, having been encouraged not just by extra CO_2 but by an increase in man-made nitrogen, this element being the nutrient which most stimulates growth in northern forests. Throughout the 20th century, in particular, nitrogen has been liberally produced by developed communities, a quantity increased several-fold in the past few decades, and more than doubling the natural quantities. Third, with forests undoubtedly important, the examination of other ecosystems, such as peat wetland, has been relatively neglected, with these regions more relevant than generally assumed. Fourth, the warming created by CO_2 may itself be stimulating forest growth, with the growing season extended by 2–3 weeks in many northern areas. There is also the little understood carbon cycle within soil where organic carbon is degraded to an inorganic form very slowly, perhaps in centuries and even in millennia. This turnover creates a 'large source of uncertainty', a succinct phrase much used by those summing up the lack of understanding.

David Schindler concluded his thoughts on a more personal note, and this summary is more than worthy of inclusion in its entirety. 'To a patient scientist, the unfolding greenhouse mystery is far more exciting than the plot of the best mystery novel. But it is slow reading, with new clues sometimes not appearing for several years. Impatience increases when one realizes that it is not the fate of some fictional character, but of our planet and species, which hangs in the balance as the great carbon mystery unfolds at a seemingly glacial pace.'

There may be a lack of haste, and little consistency, around the world concerning estimates of CO_2 and trees, with each new scientific paper likely to contradict many earlier papers, but there is considerable current enthusiasm among carbon producers for growing trees to absorb carbon, as growing trees are bound to do. A mature forest is thought to be in equilibrium, emitting about as much CO_2 as it absorbs, but young plantations absorb more than they release. Therefore the determination to replant is much like penance, a form of mortification to absolve a sin. Calculate your effluent, and then plant a sufficiency of trees to mop up that production. One Dutch power group, for example, has been financing local forestry in Sabah, Borneo, to absorb a carbon quantity precisely equivalent to the amount it was creating back home in the Netherlands. Costa Rican farmers are being paid $10 by Norwegian and American businesses for every ton of carbon their private trees are absorbing from the atmosphere. The farmers do not have to plant the timber; they must merely let it grow.

Growing trees do absorb carbon, but there is doubt about almost every other aspect of this sequestration, as it is being called. How much carbon is actually absorbed, and also released, by trees as they grow, and mature and die? How much is released should the trees be wholly or partly burnt? Does the extra nitrogen, by causing speedier growth, also lead to less nutritional value in leaves which insects eat, as has been suggested, and therefore to fewer beneficial/harmful insects? How much less carbon is absorbed if it is a dry year, and how much more when it is wetter or warmer, or both? Trees vary, according to their species, their age, their locality, their climate. Is it best if trees are actively produced, as with the wood-pulp forests of Canada and Scandinavia, before being cut and instantly replaced? And what happens to all the carbon in the cellulose these trees produce, the paper, the wood, the chipboard? Is anything to be gained for carbon absorption by letting mature trees stand, however wonderful they look, with all their gathered carbon already locked into their wood? The single certainty, as at the outset of this paragraph, is that carbon is absorbed by growing trees.

Nevertheless calculations have been made, with estimates fluctuating widely. A tropical plantation, it is said, can absorb 40 or more tons per acre in 50 years. As for a suggested global reforestation programme of 870 million acres, an area larger than the European Union, that could lead, according to the Intergovernmental Panel on Climate Change, to the sequestration of 35 billion tons of carbon in 50 years. This figure is equivalent to 6% of the expected carbon dioxide emissions during the first half of the 21st century. According to Norman Myers, already mentioned in this chapter as a tree devotee, 'Forests are fundamental to our biosphere's workings . . . A salient way to resist global warming would be through grand-scale reforestation of the humid tropics, where trees grow several times faster than in, say, Britain.'

Disarming comparisons have also been made for the next 50 years between:

(a) carrying on as normal
(b) accepting the Kyoto protocol
(c) accepting the protocol and planting trees.

It is thought that the three respective figures for atmospheric CO_2, in parts per million, would after half a century then be: (a) 550, (b) 430, and (c) 380. These totals, along with all the others, may prove to be faulty, even grossly so, but there is agreement on their general message. If the Kyoto protocol is adopted there will be less CO_2 than might otherwise exist. If Kyoto and forestation are implemented there will be even less. If nothing is done, and there is business as usual, there will certainly be more of this undoubted greenhouse gas, with the world paying a price for its laggardly attitude towards the lessening. Or rather all people of a future world will do so, the fault being ours, not theirs, but firmly theirs to reap.

Planting trees and producing less CO_2 will both be beneficial, but inadequately. Even if society were to cut back intensively – and unbelievably – on consuming fossil fuel (which today accounts for 85% of the world's commercial energy needs) there would still be a CO_2 problem, with this gas remaining in the atmosphere for a century or so. The sins of the fathers will be descending upon the great-grandchildren, or even further. Therefore plans are afoot for the more positive removal of this greenhouse gas than expecting trees to serve as mops. Just as oil is extracted from deep beneath Earth's surface so could carbon dioxide be despatched downwards as a form of replacement. According to promoters of this idea the unwelcome gas 'can be pumped into underground geologic formations, such as unminable coal beds, depleted oil or gas wells, or saline aquifers'.

It will have to be stored safely to avoid creating a tragedy similar to that of Lake Nyos, Cameroon, in 1986. A huge bubble of carbon dioxide then belched from that lake, promptly suffocating 1,700 people, but the storage promoters argue that such a mishap would not result from their proposals. Already some have been put into effect, such as the Sleipner gas rig 150 miles off the Norwegian coast which pumps CO_2 into 'a saline aquifer 1,000 meters below the sea-floor'. Many oil wells in the United States pump carbon dioxide underground, not so much as a form of sequestration but to improve the efficiency of oil drilling.

Water This colourless, odourless liquid, is, when vapourised, an undoubted greenhouse gas but that, once again, is the solitary certainty. If water vapour is abundant in the atmosphere it is the most important greenhouse gas, but its abundance is highly variable. The average quantity of atmospheric water is 3,150 cubic miles, a little less than exists in soil and slightly more than is present in marshes, but the water within air varies greatly from place to place and also with altitude. When it condenses to form cloud its properties regarding heat conservation and radiation change dramatically, quite apart from there being so many kinds of cloud.

Those 3,150 atmospheric cubic miles are impressive, but represent less than 0.001% of the water on this planet. Practically all of it (96.5%) is in the oceans. Of the fresh water, so crucial to all life forms – with human bodies being 65% water, and humans without water dying so speedily – a further 1.7% is in polar ice, another 1.7% is in groundwater going down to 2,000 or so feet below the surface, and the remaining 0.1% is either on or near the surface or in the atmosphere. Rivers contain 0.006% of the water total, or 0.02% of the fresh water total, a fact giving the lie to all those frequently repeated Amazon pronouncements about this single river containing one-fifth of the world's fresh water. Even the plants and animals now alive contain half as much water within their systems as do all the rivers.

Atmospheric moisture also maintains life by being part of the hydrologic cycle. Water is evaporated to join the atmosphere from the oceans and the land, these proportions being 87.5 and 12.5 respectively. This water then falls, as precipitation of one kind or another, on the water and the land, these proportions being 79.4 and 20.6, again respectively. Water only stays in the atmosphere for 8.2 days, on average, but it is this water vapour which acts as greenhouse gas so unpredictably. Although all such gases serve as warming agents, and water vapour is no different, the ability of water also to form clouds causes it to act in part as coolant by diminishing solar radiation. Therefore more water vapour may lead to less heat which then leads to less water vapour – no wonder there is uncertainty about the greenhousing effect caused by this most important of all the greenhouse gases.

Carbon dioxide's presence in the atmosphere is straightforward. The more there is, from whatever cause, the more heating will ensue; save, of course, when water enters the equation. More heat caused by CO_2 will create more water vapour, and therefore more cloud, and therefore less solar heat reaching Earth, and certainly more uncertainty in the minds of those who are attempting to predict the consequences of all this interaction.

Some good news abruptly surfaced in August 1999. Emissions of carbon dioxide fell during the previous year by 0.5% despite a growth of 2.5% in the world's economy. This was the first such fall since the start of the 1990s when the collapse of eastern European economies had also led to a temporary drop. Economic growth in the United States, the world's largest CO_2 emitter, had risen by 3.9%, but its emissions had increased only by 0.4%. China, second largest emitter, experienced economic growth of 7.2%, but an emission growth of 3.7%. Poland, even more encouragingly, had an economic growth of 6%, but a 9.7% cut in CO_2 emissions. Both China and Poland had made their improvements by diminishing the previous enthusiasm for coal.

These relative improvements cannot be credited to new government policies. Instead, apart from the blessing of cuts to coal consumption, and the greater favouring of oil and gas, there has been a beneficial switch in the manner of economic growth. Many modern industries, such as information technology, are far less polluting than old-style factories. Chris Flavin, of the Worldwatch Institute in Washington, DC, recently announced that the entire Internet 'uses less electricity than New York City'. There is now a wide belief, despite the United States' unwillingness to ratify the Kyoto agreements for 'fear of damaging the US economy', that the Kyoto strictures may have been too lenient. Emission cuts of 60% are thought to be needed to stabilise carbon dioxide at the current (higher than normal) levels, and even greater percentage drops must be achieved if those levels are to be reduced.

Methane – CH_4 – is created when organic substances decompose without the presence of oxygen. It was originally called marsh gas because it bubbles from swamps where dead material is rotting, as children learn when happily poking sticks into oozing foetid ponds. Natural gas, used widely as a fuel, is largely methane. The escape of this gas during the process of extraction contributes about 15% of the methane present in the atmosphere. A further 18% comes from animals, notably the domestic herds of cattle, sheep and pigs. Enteric fermentation, which leads to belching, is to blame. Other big percentages are from rice cultivation – 30% – with paddy fields more swamp-like than the fields of other crops, from clearing land for agriculture, from animal waste, from sewage farms, and from the decay of human garbage. The world's coal mines are believed to release 25 million tons of methane annually, or as much as leaks from oil and gas fields. Termites are also a major natural source, producing about one-sixth as much as swamps and marshes. Some frozen lakes of northern Siberia possess channels in their ice kept open by methane bubbling from peat sediments at their base.

The two most worrying methane facts are that its atmospheric level is rising at almost 1% a year, twice the speed of CO_2 increase, and the gas is many times more effective per molecule than is carbon dioxide in swelling the greenhouse effect. Methane's current contribution to that effect is about 20%. Human activities produce twice as much methane as is created naturally; hence the doubling of its level in the atmosphere since pre-industrial times and its present rise. As an increase in rice production – some say by 60% – is the most appropriate way to feed the population increase of the next 30 years it is highly likely that more methane will be created to satisfy human appetite. Currently rice is the staple food for two billion people, with Asia producing 90% of it. In 1990 paddy cultivation was being blamed for over 100 million tons of methane production per year, a figure – according to E.S.R. Gopal, of Bangalore – now much reduced with earlier calculations being blamed for the error.

Some good, and contrary, news is that the rate of atmospheric methane increase has actually been decreasing recently, although the reasons are not known. It is even postulated that, during the next 20 years, its level will only be a few per cent higher than now and may then reach a plateau. This possible levelling is despite the fact that human numbers will almost double before they reach a similar plane (including many hundreds of millions more rice-eaters), and that natural gas will increasingly be favoured over coal and oil as a source of energy, this methane-rich hydrocarbon sometimes leaking to the atmosphere. Methane's lifetime in the atmosphere is about ten years, but this period can be reduced by OH, the hydroxil radical. The quantity of OH is itself affected by carbon monoxide and by oxides of nitrogen. These three gases are all produced by the internal combustion engine. Therefore, although methane is unwelcome for its greenhouse properties, the gases which destroy it might be considered welcome, save that CO is a poison and NO_2 is also damaging to life.

China's economy is developing fast, possibly faster than ever before, and China's power (for, let it never be forgotten, one quarter of the world's population) comes mainly from coal-fired plants which belch forth large amounts of sulphur dioxide. Russia's economy is developing much less fast, but is also SO_2 guilty. For example, much of the platinum and other metals used in catalytic converters for cars comes from Russia. Such converters, now mandatory on new cars in most of the industrialised world, do indeed reduce unwelcome hydrocarbon pollution by 90% but, according to some German researchers, the Russian refining plants produce over six times as much SO_2 as equivalent Canadian enterprises. Globally the platinum producers therefore undo much of the good being achieved by the converters.

Geological events tend to have timescales measured in millions of years rather than thousands, but there was a burst of methane gas some 55 million years ago which is thought to have raised deep ocean temperatures by some six degrees C. within 10,000 years. This is the kind of haste which certain scientists are predicting for the effects of human activity. Examination of the sediment off Florida's north-eastern coast has shown the loss of over half of the foraminifera in that geological eye's blink of one hundred centuries. A sudden release of methane is thought to have been responsible for the abrupt slaughter of these bottom-living creatures. They could not stand the heat.

Sulphur dioxide is a greenhouse gas, and therefore affects the climate, but it is also the principal component of acid rain. Killing off trees may indirectly affect the weather, and it is additionally disturbing that rain itself, the 'soft refreshing rain . . . sent from Heaven above', can be unfit to drink. Much of Europe's precipitation, according to a Swiss study, contains such high levels of pesticides that it would be illegal to supply as drinking water. Medical substances are also reaching high levels in water, not so much because old medicines tend to be flushed down the toilet but because we excrete them in our urine. Human activities are so numerous, so intertwined and so frequently damaging, that sorting the good from the bad, and deciding which is dominant, is fearsomely complex.

In similar fashion the greenhouse effect is far from being a clear-cut event. Neither are the chemical processes in the atmosphere which either mitigate or bolster it. What is entirely clear is that greenhousing, ozone depletion and a changing atmosphere are all planetary problems. The Montreal Protocol was praised because so many nations acted so promptly in unison. Nations will have to act increasingly in unison for the elementary reason that they all exist on the same planet. A hurricane, typhoon or earthquake may lay waste one particular community, but increasing levels of CO_2 or CFC or methane affect the planet as a whole. The history of mankind has always been a local matter, with each group/tribe/region/nation either belligerently or peacefully obsessed with its own requirements, being us against them, our kind in preference to their kind, Houyhnhnms against Yahoos. In future the human race will have to be a great deal more global in its thinking, and less parochial, if there is to be a future history.

Chapter 14

ALTERNATIVE ENERGY

Oil and hydrocarbons – coal – wind power and wind farms – wave energy – tidal currents – nuclear reaction – hydropower

A botanist was standing in the open, his face and body inclined towards the sun. 'Why can't I be like the trees?' he lamented; 'Why can't I stand as they do, and get energy from some kind of photosynthesis?' He had a point. The process would be so simple and so clean. (It would also be extraordinarily tricky, with the process of splitting water to form oxygen no trifling feat.) Human efforts to acquire power have, in general, been irrevocably accompanied by some unwelcome and polluting attribute. Burn coal, and there are all kinds of noxious effluents. Burn oil, and the noxiousness is merely different. Burn natural gas, and methane escapes to harm the atmosphere. Exploit nuclear fission, and radioactive residue will last for centuries. Dam rivers, and great tracts of land are drowned. Harness the wind, and huge propeller blades noisily litter lonely areas. Everything has a price, a sum which must be paid.

Apart from such primary disadvantages there are secondary snags as additional discouragement. The combustion of any hydrocarbon creates great quantities of carbon dioxide. This is a greenhouse gas, not only warming Earth but creating an increasing demand for the kind of power which cools. Nuclear energy involves a taming of massive energies, a taming which can go so very wrong when wildness takes its place, when control is lost and mayhem reigns instead. Every blocking of a river leads to problems with former freedoms, concerning fish, fishermen and local life in general. The Three Gorges dam being built on China's Yangtze River will create a reservoir 375 miles long and displace 1.4 million people.

Wind power is best achieved when strong winds are blowing, but humans want power as and when they choose. Nuclear reactors function best when working steadily, but their great energy must be stored when demand for it is slack. Humans want power where they live, but where they live is rarely the optimum place for power generation. Therefore energy must be moved, disfiguringly and expensively, via transmission lines for miles and miles and miles. There is no such thing as a free lunch, say those who give and eat them. There is no easy way to manufacture power, say all of us increasingly, having steadily learned of the many drawbacks to partner all our gains.

Demand for power is growing remorselessly, both in developed and developing areas. Each year more and more contrivances become available which feed off electricity, more labour-saving creations, more creations we never knew we needed. In 1940 the world was being told that oil reserves would expire by 1970. They are not expiring even after another 30 years, but will do so some day. Almost a trillion barrels of oil have so far been consumed, and perhaps another trillion are still down there, sufficient for a further 40 years or so. Therefore, from climate's point of view, there is an awful lot of carbon dioxide still to be released from its age-old imprisonment within those hydrocarbons.

Concern will undoubtedly increase about the side effects of power generation, notably in countries sufficiently wealthy to afford such concern. Renewable resources will more frequently be favoured, whether or not the electricity they generate is more expensive. Intermediate technology will help developing areas, with everyone aware that deforestation and primitive methods are no longer so acceptable. Traditional fireplaces, oozing smoke continually and forever damaging to lungs and eyes, will no longer be seen as picturesque or even admirable, any more than Birmingham's 280 satanic chimneys which were so proudly featured and upstanding one century ago. Power will always be needed, but its generation will no longer be a matter of cost alone. What it is doing to health, to the environment, to the climate and to all our descendants who are next in line will become no less critical.

Coal was king when the 19th century became the 20th. It had taken the place of windmills, of water wheels, of burning wood, of horse and human power. By no means has it abdicated, but its throne is toppling. Nuclear power did not exist as a source of electricity until Britain's young Queen Elizabeth pressed the button to start up Calder Hall in 1956. Such reactors now generate 17% of the world's electricity, a figure likely to rise, despite Calder Hall's accident and fire one year after that official opening and despite Chernobyl's far greater horror in 1986. Nuclear stations produce 30% of Britain's electricity, 50% of Scotland's, and 60% of France's. The radioactivity they also create, whether bulky and of low activity or more compact and highly active, is a massive problem, but at least that global

Alternative Energy

17% of nuclear power is preventing the emission of over two billion tons of carbon dioxide every single year.

The longing for a renewable source of pollution-free power has boosted research into alternative possibilities. This was unthinkable and presumed quite unnecessary a century ago, when coal was pre-eminent and almost half a million Britons burrowed mole-like along the seams to excavate this vital raw material (laid down mainly in the Carboniferous period of the Palaeozoic era about 300 million years ago). Take wind power, for example. It was important when the British Isles possessed almost 10,000 of the traditional four-bladed sail windmills, with dusty millers making flour from corn, and with the very few surviving structures now revered as enchanting relics from those simpler wind-powered days. When coal came along to displace the creaking, cranking and whirring mills, so picturesque but so stationary when winds failed to blow, they were quietly neglected and their millers abandoned them for work elsewhere.

Wind will become important once again if the British Wind Energy Association has its way. The United Kingdom, always a breezy spot as first European item to be encountered by Atlantic storms, possesses over half the continent's total wind resource. The BWEA has advocated that 6% of Britain's energy demand in 2010 should be met by several thousand offshore and land-based turbines. The Countryside Agency, among others, is disturbed that wind farms would be located in some of the country's most precious places. 'An ambitious programme of wind energy,' said the Agency's Head of Planning, 'will mean we give up our wildest countryside for a very small energy contribution.'

The first such British farm was commissioned in 1991. It had a capacity of four megawatts, as against the 2,000 MW produced by many a modern power plant. By 1997 some 40 wind farms had been installed in Britain, their capacity totalling 200 MW, with a further 137 MW 'on line'. Such a figure, although an impressive gain from nothing at all, is still far short of the power from a single, conventional, CO_2-producing, coal-fired station, with coal continuing to provide 90% of the fuel for all non-nuclear steam plant. Wind is entirely non-polluting, save for being visually and audibly disturbing. It provides an everlasting supply and is totally renewable. It is reasonably inexpensive and satisfies many of the criteria being demanded by increasingly vociferous pressure groups. Some home-owners, as on the windy Isle of Man, have even installed their own personal wind generators, which cut down on electrical units taken from the mains.

There are also disadvantages to counter optimism. The turbines work most efficiently where there are steady winds, as in the trade-wind belts.

They work less well in Britain where wind can be non-existent or blowing a gale. The mills do not turn until wind speed is sufficient, perhaps 15 miles an hour. They also do not operate when wind speed is excessive, for fear of overheating. Consequently their electrical output is inconsistent and certainly bears no relation to the times of human need. One Cornish contributor to *New Scientist*, whose home was near a windmill 'farm', concluded his diatribe: 'They are not pretty, not quiet, not useful.' With impartial editorial juxtaposition his letter was followed by another stating that wind energy could save 750,000 tons of carbon emission out of the 36 million target set by the government before 2010. 'The thousands of wind turbines . . . will be absorbed into our landscape in the same way as the 10,000 wind turbines that operated throughout England in past centuries.'

Windmills were, of course, erected without recourse to any organisation, such as the planning inspectorate or the Countryside Agency. At the end of 1999 the High Court decided that a planning inspector was right to forbid the construction of what would have been England's biggest wind farm. National Wind Power had sought permission to erect 25 turbines, each 177 feet high, in County Durham near Barnard Castle. The site may have been suitable for wind power but it was also on the border of both a national park (the Yorkshire Dales) and an Area of Outstanding Natural Beauty (the North Pennines). The 'discordant visual impact' was held to be sufficiently unwelcome, and the gain in electrical energy sufficiently small, for this application to be refused.

The United States is waking up to the benefits of wind power, with the Clinton administration in July 1999 setting a goal of 5% of the nation's power needs by 2010, an advance of 4.9% over the 1999 levels. 'Wind energy has been the fastest growing source of energy during the past decade,' said the energy secretary while pointing out that wind power worldwide exceeded 10,000 MW. The American Wind Energy Association, in welcoming the government's statement, called it 'the most visible and aggressive commitment for 20 years'. It is claimed that wind-generated electricity, the closest in the renewable energy field to being commercially competitive, costs five cents per kilowatt-hour, an eightfold drop since 1980. Farmers would consider wind power 'a new cash crop', and allegedly there would be little objection to the new technology as 'most installations' would be erected on 'remote agricultural land'.

By the end of the 20th century Europe had surpassed even the wind industry's own targets. In 1999 alone wind-generated electricity rose by 30%, mainly due to expansion in Germany, Spain, Denmark, Holland and Italy. According to the European Wind Energy Association this growth meant that wind capacity had increased to 8,900 megawatts – equivalent to the output from four major power stations, whether conventional or nuclear. The EWEA believes that capacity will be 100,000 megawatts by 2020, thus supplying about 10% of Europe's needs.

As for wave power, Britain is not only well supplied with waves but also with ideas for making them yield a portion of their energy. The most determined wave proponents, such as Greenpeace and British wave consultants to the European Union, say 'half' of Britain's electricity requirements could come from waves 'by 2040'. The British government, via its principal wave adviser, is less fulsome but still optimistic, believing that 20,000 MW could be tapped from British waves. They would therefore satisfy 20% of British power needs. He also considered that, following further research and development, wave power could become commercially competitive.

Currently not even a single megawatt is being produced but, as of the end of 1998, a total of 15 'wave power generators' were being actively planned, the majority in Europe, with these developments much heralded at the Third European Wave Energy Conference held in Greece that October. Unlike every other form of power generation there is little price to pay beyond that of the equipment and its maintenance. There is certainly no effluent and, unlike wind generators, no whirring and obtrusive structures offending precious bits of countryside. Instead there are constructions, half on land and half in water, which, in most designs, do not move externally but have internal moving parts.

In essence the incoming waves first expel air from, and then inhale air within, a chamber. In going back and forth these quantities of air pass through a turbine and, most ingeniously, the turbine's blades rotate in the same direction whichever way the air is moving. Therefore no valves, and no stopping and starting of the blades. As of now it is expected that such a system will be able to extract 40% of the power from that moving air, and this figure could well be made to rise. There are also schemes for variable pitch turbines, a possible – but as yet unproven – improvement in wave energy extraction.

Inevitably, as with the wind machines in wild and empty areas, there will be resentment if precious regions have to be dedicated, via major pieces of machinery, to the harnessing of power. After all, there is only one foot of shoreline for every British citizen, even if the several hundred islands are brought into the equation. It will not merely be more pleasing for all those inhabitants if the generators could be located offshore, but more satisfactory. The wave energy out there is greater, with more power to be tapped.

Several schemes are being planned for deeper water, with all manner of floating tubes, oscillating buoys, moving rafts and 'nodding floats' involved. Estimates have been made that offshore floats might generate power at 2.4 pence per kilowatt-hour, as against the 4 pence per kw/hr for new coal or nuclear stations. Plainly the development of wave power, still at an infant stage, is likely to proceed apace, with no one knowing which scheme will become most satisfactory. (To an outsider all the concepts, and the several forms of optimism, can seem much like aircraft design in, say, 1909, when there were rudders in front, rudders behind, wheels, skids, pusher propellers, two wings, ten wings, rotating wings and warping wings, with no one having any real idea which kind might prove most sensible.)

It can be argued that the steadfast talk about global warming, with its varied threats of inundated land, excessive heat and destabilised weather, is already serving as a blessing for wave and wind research by encouraging the development of less offensive forms of power generation. In 1997 Denmark launched the first phase of its national plan to create 5,500 MW by wind power before 2030, this total being half its national consumption. Germany plans to build the two biggest wind farms, with 200 mills some 20 miles off Rügen island in the Baltic and another 200 in the sea off Helgoland. Norway intends to build the world's first tidal mill power station, entailing the sinking of turbines within the strong tidal currents of Kvall Sound.

On land there are plans for hybrid delivery vans, fitted with electric motors *and* conventional engines. The dual arrangement is hoped to provide greater fuel efficiency for such stop/start vehicles. There is even a scheme to build a solar chimney in South Africa, a funnel one mile high and over four miles wide at its base which, if fact matches theory and the thing can actually be built, would generate 200 MW by solar power alone.

As from April 1998 British home-owners were confusingly being solicited to 'buy cheaper gas from their electricity company' and vice versa with the electricity companies offering gas. This resulted from the former monopolies losing their exclusive rights, and the subsequent, if complicated, freedom did much to boost thoughts about alternate means of energy acquisition. In that year a mere 2% of Britain's electricity was generated from renewable resources, such as hydropower. The government then pledged to increase this quantity to 10% by 2010. An official study, 'leaked' in August 1998, asserted that renewable energy could generate 'half' of Britain's electricity by 2025. This renewable half would allegedly cost less to produce than electricity from new coal-fired or nuclear stations, and would entail new systems. A problem with hydro-electricity, although clean and involving a renewable source, is that most sites in developed nations have been almost fully exploited, since these are so dependent upon geographical availability.

The world's first nuclear energy plant contributing to the national demand for electricity opened in Britain's Cumbria during 1956. Admittedly the driving force behind this dawn of a new age was the demand for a substantial quantity of plutonium, the critical ingredient of one form of nuclear bomb. Not only was Calder Hall able to produce the bomb's raw material but it donated a major measure of electricity to the national grid. The future then looked bright for nuclear power.

Soon afterwards, in those 1950s and 1960s, Britain started building other and entirely civil nuclear stations, as at Bradwell, Hinkley Point, Trawsfynydd, Hunterston, Berkeley, Dungeness and Sizewell. Simultaneously there was much optimism at Dounreay, where more nuclear fuel would be produced in a novel 'breeder' reactor than was consumed, and at Windscale where a new kind of power generator, the AGR, was expected to be more efficient. There was also excitement at Harwell where work was well advanced with a device called ZETA. This was intended to acquire energy, as with thermo-nuclear bombs, from the heavy form of hydrogen, this deuterium being freely available in ocean water. There was even the prospect of using nuclear reactors to power conventional cargo ships.

Since then, as the world knows, the ebullience over this brand-new form of energy has been somewhat muted. The disposal of nuclear waste for thousands of years, the fear – and then the reality – of accidents (starting even with Calder Hall/Windscale in 1957 and on through Idaho Falls and Three Mile Island to Chernobyl) helped to diminish lingering enthusiasm. So too did the increasing unhappiness about living near nuclear power plants. None of this warmed the population in general to the existence and development of nuclear power.

Nevertheless nuclear stations have prevented the emission of enormous quantities of CO_2 by their very existence as suppliers of energy. Their protagonists say they have cut this contribution by 8% a year. Therefore, as a spokesman for the industry recently proclaimed, they have 'contributed substantially to meeting the sustainable energy challenge ... [Nuclear power] is the only readily and commercially available "no carbon" electricity generating option other than hydropower which makes a positive contribution to a reduction in CO_2 emissions.'

In truth there are many processes involved in nuclear power production which consume energy and produce CO_2. Nuclear stations are therefore not zero emitters, but their output of the gas relative to hydrocarbon consumers is minute. Even though most industrialised communities create some of their energy requirements by nuclear means, there is no longer the joy felt in 1956 at Calder Hall when Britain's Queen pressed the start-up button. She was then 30, and the nuclear industry was even younger, with 1942 being the year when an atomic reactor first started work. It has aged a lot since then.

Currently coal, oil and gas are the powerful triumvirate producing most of the world's power, but they cannot last for ever and, with global warming already happening, will be increasingly resented with the kind of antagonism currently directed towards nuclear reactors. Every one billion kilowatt-hours of electricity generated by renewable means will reduce carbon emissions by 200,000 tons, sulphur emissions by 10,000 tons and particulate matter by 2,000 tons. Even if renewable energy does cost more, as is highly likely at the start, it may increasingly be considered good value for the greater price. It may even be thought of as the only proper way to generate the energy on which life depends, much as many earlier and degrading systems – child labour, partial suffrage, discrimination in all its forms – are being abandoned, having no place in a modern world.

Early in 2000 Britain's prime minister, Tony Blair, stated that British electricity companies must be supplying 10% of their power from renewable sources within ten years. This should reduce, by some 2%, Britain's annual emissions of carbon into the atmosphere. This lessening of greenhouse gas is therefore modest; moreover, by placing responsibility for change with the generating companies, these will inevitably use the cheapest means for acquiring that 10%. Wave power and offshore wind power, both currently more expensive than land-based wind power, will assuredly be disregarded. Therefore the governmental stricture, however high-sounding and seemingly laudable, will be barely influential on the global problem and not much of a stimulant for research into future forms of renewable and non-polluting power. Perhaps all governments are inevitably weak in seeing beyond the next couple of elections.

To sum up. The world's alternative means of generating energy – solar power, wave motion, wind farms, tidal flow – are all intriguing but largely irrelevant for the near future. Their contribution to the demand for energy will almost certainly be small in the years immediately ahead. Nuclear power is a force to be reckoned with, but no longer the great white hope it used to be. Hydropower has already creamed off the best places.

Amazonia has tremendous hydro-potential, but only at equally tremendous cost to the local people, to their land and nature. Most of Brazil is extremely flat, and therefore not readily conducive to hydro-power, although dams have been built within the drainage basin of Amazonia. The massive Tucurui structure on the River Tocantins, with the world's largest spillway to cope with the rainy season's annual surge, generates huge quantities of electricity, but it created a lake some 200 miles long and extremely wide. The land beneath has therefore disappeared.

Hydrocarbons will continue to be the mainstay of power generation. Therefore carbon dioxide, however much its output is trimmed here and there, will continue to be emitted in tremendous quantities. Its effect as a greenhouse gas will become more and more important. Alternative possibilities for power will be developed, and will help, but CO_2 levels will rise even if, with a magic stroke, the burning of coal, oil and gas for electrical generation could be cancelled overnight. People are increasing, and people, whatever their wealth and form of livelihood, will always add to the quantity of the most influential greenhouse gas of all.

Chapter 15

CLIMATE VARIATION

Freak years – climate modelling – recent change – ice – glacial retreat – sea level – shoreline adjustment – doing something, doing nothing – inundation – global impact – Babel of conflict – the net – computers and forewarning

On 20 July 1911 'furious rioting' broke out in South Wales, an explosion of workers' discontent reported extensively by British newspapers. Nine people were killed, three of them by soldiers' bullets. On 8 August that same year 50,000 British troops were organised to quell other rioting. Two men were then shot dead in Liverpool after 200,000 stevedores, railwaymen, carters and other individuals sympathetic to their cause had taken to the streets. ('Justifiable homicide,' concluded an inquest.) 'War declared! Strike for Liberty!' proclaimed the dockers' placards. James Keir Hardie, Labour leader and MP, addressed strikers at London's Tower Hill: 'The masters show you no mercy. They starve you, they sweat you, they oppress you. Pay them back in their own coin.'

Coincidentally, and possibly helping to stoke the situation, that year witnessed the hottest summer Britain had known for a very long time, with numerous long-standing records being vanquished, along with all those lives. During the four months of 1911 from June to September there were 15 weeks in Britain's southerly half where temperatures were substantially above normal. Even in early July it was extremely hot, with Greenwich registering 88 degrees Fahrenheit and Epsom 90 degrees (31 and 32 Centigrade). By the end of that month the cities of Bath and Rugby were recording 94, Greenwich was sweltering at 96, and Epsom at 97. August then proved to be hotter still. 'Various parts' of London recorded 97, Canterbury and Epsom reached 98, Isleworth 99, and the Royal Observatory at Greenwich 100 degrees, this being higher than any temperature it had ever measured on any day since regular recordings had begun there in 1841.

That maximum heat was equal to 37.7 degrees Centigrade, or a modest fever in human terms. Small wonder that Thomas Burgess chose that sweltering summer to swim across the Channel on 6 September, the first person for 36 years to do so after Matthew Webb had been the pioneer, and therefore only the second such swimmer in recorded history. July 1911 was extremely dry as well as hot, this dryness never exceeded since that time until the July drought of 1999, another torrid month.

The severe heat of 1911 took its toll on all the population, most notably its children. During July's final week it was reported that 382 of them had died nationwide. Towards the end of August that figure, for children under two, had shot up to 855 within a single week. London later revealed that a total of 2,500 infants had perished during the heat wave, with the capital being named by newspapers as the 'second most unhealthy city' in the world (without mention of the unhealthiest – Naples? Bombay? Glasgow?). London's mortality rate, for all ages, was then reckoned to be 19 per 1,000 per year, as against today's comparable figure nearer to 10 per 1,000. Americans were also suffering during that exceptional summer, with 652 reported dead from heatstroke during a single week.

September 1911 was not quite so hot, but still remarkable. On the 2nd of that month much of southern England exceeded 85 degrees, with 89 at Greenwich, 90 at Camden Square and Cromer, and 91 at East Ham. A few days later it became hotter still, with thermometers soaring to 90 almost everywhere in Britain's lower half. On 8 September the maximum recorded at Hampstead, Cambridge, Rugby and Bath was 93 degrees, while Isleworth and Greenwich were one degree hotter. Records were therefore still being broken, with the Royal Observatory recording the highest it had ever registered for any September day.

Later that year, and after the country had cooled to more traditional temperatures, no one stated either that the strikers' grievances might have been aggravated by the searing heat or that a few of the soldiers' bullets might have been loosed for the self-same reason. And certainly no one mentioned that the extraordinary summer was perhaps one indication of the planet itself becoming hotter. Or that the filth pouring from all the nation's industrial chimneys might be in part to blame. Or that a country powered almost exclusively by coal was belching carbon dioxide into the atmosphere at an unprecedented rate.

Eighty-eight years later, during the summer of 1999 and in Moscow, there was a similar heat wave in a city more accustomed to climatic hardships during winter. No one officially blamed global warming, with countless Muscovites more ready to blame each other for misdeeds. Several murders were entirely heat-related, with one man lethally incarcerating his wife in a refrigerator 'to cool off', and another bludgeoning his spouse to death because she had continued to demand air-conditioning instead of the fan he had acquired. (In any case murders are now three times more common in Russia even than in the gun-owning, and so frequently gun-using, United States.) It was the hottest summer Moscow had known for at least a century, with temperatures well into the 90s. Countless citizens flocked to pools, to reservoirs and to rivers for relief, whatever the condition, or danger, of the open water. The presence of 'No Swimming' signboards merely provided confirmation that swimming was feasible in that area. During June a total of 140 Moscow citizens drowned in their enthusiasm to find respite from the heat, with alcohol encouraging their immersion and a frequent inability to swim proving insufficient as deterrent. Even nearby peat started to smoulder, as if in sympathy, beneath its covering of soil.

What Britons were experiencing in 1911 and Russians in 1999 were undoubtedly exceptional years, being many degrees hotter than normal. It is therefore correct that such temperatures were not immediately attributed to global warming, even if the phrase had not surfaced in the century's first half. There have always been outstanding exceptions, and always will be. Cherrapunji, famed even in India for massive annual downpours of several feet, can occasionally experience twice as much. Deserts, accustomed to nothing at all save unrelenting sunshine, can suddenly be deluged, with desiccated plants and seeds bursting into life, their undying optimism eventually rewarded.

Besides, has there ever been a countryman, in any time or place, who has not railed against the weather for being exceptional? Has there ever been a human in all of history unaware that climate, so often and so irritatingly, is a moveable feast, the days being hotter, colder, drier, wetter, windier or calmer than some perfect year, enshrined in memory if not in truth, when rainfall, warmth and wind all behaved correctly? Such a year has, in all certainty, never happened. All years are odd, to some degree, and merely add their differing measurements to make the average, an average which never happens but all of us remember, and all of us expect, much like childhood days allegedly bathed in unending sunshine. Or blanketed with bright snow, perfect for snowmen. And with every season behaving as it should.

Those Londoners, Welsh and Liverpudlians of 1911, along with the Muscovites of 1999, were suffering summers at least ten degrees C. hotter than they had every right to expect. Global warming, on the other hand, even for those whose estimates are at the most extreme, is being reckoned only in fractions of one degree. If some community's average July heat is 20 degrees C. or is 10 degrees C. the global warming about which there has been such anxiety and such talk has merely pushed those averages to, say, 20.6 or 10.6 degrees. They are not increments making it under-standable – let alone acceptable! – that men lock wives in refrigerators or rioting workers become victims of rifle fire from sweltering military.

The freak years of 1911 and 1999 were aberrant, and therefore memorable. We humans notice weather only when it is odd. If it is 'normal for the time of year', the phrase so favoured by forecasters, we take very little interest. So too if the week, or month, or year is only marginally different, not quite so warm or cold or wet as customary, that commonality being an amalgam of previous years ever since, as we are so often told, 'records began'. There is no way for ordinary humans to comprehend that the second week in November, for example, has been slightly drier, on average, than over the past 120 years. Besides, human beings are so subjective. Rainfall at night is of less concern, but no different for rain-gauges. Light rain is usually as annoying and thought-provoking as slightly heavier rain, but the gauges measure quantity and, in the main, record neither lightness nor heaviness. Every drop of rain at weekends or during holidays is more disturbing for humans than rain in working hours but, once more, gauges will tend to contradict personal assessments of a fallen quantity.

Forever ready to exaggerate the human ability to misjudge any situation, newspapers and television are each determinedly interested in exceptions rather than the rule. They omit that hurricanes have failed to happen recently or that one week's snowfall is exactly average. With current attention focused so markedly on the climate, we viewers and readers get to believe that typhoons must be worsening, that El Niño is being more *fuerte* and *difícil* not only for South Americans but for the rest of us, and the world is becoming far more dangerous than it ever used to be. With television reporters able to reach almost everywhere so speedily, and disasters great for pictures, it is only too easy for audiences to *know* the world is growing more turbulent, almost by the hour, certainly by the year. News from Krakatau's devastation over a century ago took seven months to reach *The Times* of London, long after its dust had darkened the European sky. News about Chernobyl's explosion arrived even before its radioactivity was quietly landing on foreign fields, along with the showers of rain which were delivering this damaging cargo.

Modelling Science ought to be the guiding light, but scientists can be

misleading, often grinding axes or vigorously expressing disagreement
with rivals in their field. Just as every oil spill is plaintively exploited by
those whose income profits from conspicuous mishap, so do many
scientists welcome or abhor a piece of news which makes next year's
funding either easier or harder to achieve. Climate modelling has never
before been so much in vogue, or so well financed as now. A vast range of
weather information is fed into computers and then adjusted, hopefully to
predict the future, with a little more rain here, more CO_2 there, less ozone
in the stratosphere, more at ground level, and more or less of everything.
The computer then prints what may, and could, or should happen in those
changing circumstances.

A team within Australia's Division of Atmospheric Research (of
CSIRO) decided, in contrary fashion, to refrain from introducing novelty
into their climatic modelling. In particular they chose to assume that future
levels of carbon dioxide in the atmosphere would remain constant. As CO_2
levels are certainly rising, and will surely rise much more, they do provide
a basis for much thinking about future warmth. The Australian approach
was certainly different, and produced some unexpected answers. Instead
of the stability which might have been assumed, once the most important
variable had been removed, the print-outs showed a turbulent and
changing climate. Severe droughts, sudden inundations, hot spots and
freakish cold were all forecast for the various regions being examined.
'Fifty per cent of the globe seems to have a 10-year drying or wetting
sequence within a 1,000-year period,' said one of the modellers when
announcing the results. Not one of each region's simulated droughts lasted
more than 30 years – which sounded like good news for Perth in Western
Australia where rainfall has lessened by more than a third in the past three
decades. Local and even global climate is extremely lively, according to
this Australian research, whether or not humans are adding to that
liveliness.

As further complication the sun itself, provider of all but a minute
fraction of our heat, may have been responsible for much of the recent
global warming. The sun's magnetic field not only affects the amount of
energy emitted but also Earth's own magnetic field, a component which
has been properly monitored since 1868. Armed with this terrestrial
information it has been possible to assess the sun's energy output for the
past 130 years. Mike Lockwood, of the Rutherford Appleton Laboratory
near Oxford, believes 'about half' of recent global warming may be due to
increased output from the sun. This estimate sounds like good news for
those who disparage the effects of increased greenhouse gases, but

Lockwood and his team believe the sun was only to blame for increased warming between 1860 and 1930. Since then, and particularly since 1970, the increasing man-made greenhouse gases have been steadily more influential.

Contradiction about human involvement in global warming is mirrored by official reaction to the changing situation. When July 1998 had been confirmed by climatologists as the hottest month ever recorded in global history the US Vice-President, Al Gore, was quick to react: 'Scientists say we are warming the planet and, unless we act, we can expect even more extreme weather – more heat waves, more flooding, more powerful storms, and more drought.' The Clinton administration is therefore attempting to persuade Congress to implement a $6 billion programme of research and tax cuts aimed at cutting down American gas emissions. 'How long is it going to take until people in the Congress get the message?' added the Vice-President.

Simultaneously that same administration of Clinton and Gore was presiding over what has been called 'the most massive expansion of oil exploration' since oil first flowed through the Trans-Alaska Pipeline in the late 1970s. Two million hectares of Alaska's National Petroleum Reserve have been opened up for drilling, this area being one-fifth of that Alaskan reserve established in 1923 for use in a 'national emergency'. There is no pressing need for the oil – currently there is a worldwide glut – but there was a desire to reduce some government subsidies benefiting the oil industry. A deal had therefore been done, one entirely out of phase with that alleged wish to 'act' responsibly and reduce the 'expectation' of worsening weather.

No wonder that ordinary individuals, aware of disagreement within the scientific community, and equally aware of governmental pronouncements conflicting with governmental actions, can be confused. This bewilderment is aggravated by the relentless assortment of climatic news, good, bad, startling, depressing, horrifying, chilling, informative, and true or false which assaults us with hurricane intensity on a scale never before paralleled since – well – since records began.

Recent change The scientific journal *Nature* was a great deal more homely in its earlier days. In 1899 one reader was casually recommending a resumption of 'South Polar research' owing to the 'unusual amount of drift-ice which has broken away from the main mass'. Someone must have taken note, or had a similar thought, as the following decade witnessed considerable research into that area, notably en route to the South Pole.

Another reader, equally long ago, thought it 'worth noting that apple-blossom had been gathered [during November 1898] in the neighbour-hood of Exeter . . . More remarkably a second crop of apples had made fair progress . . . Two of these, now somewhat shrivelled, are enclosed.' Whatever the editor thought about the gift, probably yet more wrinkled on arrival than despatch, he did not use either story as warning of climate alteration, as premonition of unwelcome change. Such thinking was not yet on the agenda.

Today, if apple trees near Exeter, or indeed anywhere, behave bizarrely, or south polar ice comes further north than usual, or any weather happening is intriguingly abnormal, this event is promptly filed under climatic change. Nothing can happen, or so it now would seem, as a chance event, or merely for reaching an extreme point towards one or other end of the normal range. Or even further than that ordinary extremity, a set of strangenesses having combined to create some extra strangeness which had never previously occurred. Why should not the occasional apple tree, such as the Red Quarander of that Devon story, not respond in unique fashion to what was probably a unique set of circumstances, including one *Nature* reader wishing to inform? Similarly any hint of ice alteration these days is immediately linked to its possible association with climate change and global warmth.

Ice is a good item with which to start this assessment of Earth's warming. The polar regions have been called 'highly sensitive' to climate alteration, with temperature change more dramatic and conspicuous than elsewhere. A one degree warming may be sufficient to melt some ice, and therefore to expose more land or water. This exposed area will then receive more solar radiation as a result of its altered albedo, and will therefore experience a further gain in temperature, a further loss of ice, and further solar warmth. Consequently the polar situation is self-enhancing, whereas a one degree warming within the tropics will do nothing of the sort. It may boost growth a little, and may create a little more atmospheric moisture, but the resulting increase in cloudiness might even counter that gain in temperature by diminishing the arrival of solar heat. Ice is like a probe or an insect's antenna, giving early warning. A worker on the Antarctic peninsula, where temperatures have risen higher than elsewhere on the continent, has said the region 'could turn out to be the canary in the mine'. On the other hand its temperatures have been recorded for less than a century. Ice shelves may have been cracking up for millennia while no one was around to notice.

That altered albedo is less straightforward than might be supposed. It is true that ice reflects heat more than a similar area of exposed ground. Therefore such ground ought to be warmer than ice. But, when snow is melting, the large amount of energy required to melt it causes a cooling of the immediate atmosphere. Moreover the melted snow or ice contributes to soil moisture, and this – yet again – alters the energy balance. Any heating of the exposed ground is therefore delayed owing both to the wetter ground, caused by the melting, and to the energy required in evaporating some of this moisture. For this intriguing and supplementary confusion to the problem of albedo I am indebted to a 1998 review of 'European Snow Cover' in the *Geographical Journal* by úna nì Chaoimh (a name most memorable – or not).

Now to some wholesale statistics. The Arctic lost 13,000 square miles of sea-ice per year between 1978 and 1996, according to NASA's Space Flight Center in Maryland. This represents 2.8% of the total quantity of northern ice per decade, but the loss was not uniform either within or near the Arctic Circle. Ice actually increased in the Gulf of St Lawrence while lessening by 10.5% in the Kara and Barents seas. Not only was the total amount of sea-ice shrinking but also its depth. Measurements made by submarines indicate that the Arctic ice above them thinned by 12–15% in the 20 years after 1976. During the 1970s its depth had averaged about 19 feet, but this then shrank to 15 feet, or even less, in several areas.

As for the opposing Antarctic some satellite pictures of 1998 revealed that 78 square miles of the continent's northernmost shelf had vanished during the southern (Austral) summer of 1997/8. This probably means, reported a glaciologist, the 'beginning of the end' for the 6,000-square-mile shelf, now disintegrating steadily. This cold slab may be the smaller of the continent's two enormous shelves, but it still contains 780,000 cubic miles of ice. Antarctica's peninsula, the continent's northward-jutting extremity pointing at southernmost America, has lost 3,000 square miles from its ice shelves in the past 50 years, with local temperatures rising in that same half century by 2.5 degrees C.

An intriguing form of confirmation for this trend is written in whalers' records. These show that their vessels had to travel further and further south each year to encounter their mammalian quarry which was feeding off krill by the ice's edge. In 1930 they were voyaging to latitude 61.5 South, but to 64.3 South in the 1980s, by which time whale stocks had been depleted and a hunting ban was being instituted. The whole of the West Antarctic ice sheet will be gone in the next 1,000 to 5,000 years, according to a NASA geophysicist.

The general sea level rise, currently running at 2 millimetres/0.08 inches a year, can be accounted for in part by the melting of ice from Antarctic shelves. West Antarctica's ice sheet still contains 920,000 cubic miles of ice. If all this were to melt the world's sea level would rise by 13–20 feet, an increase inevitably causing considerable flooding of low-lying land. Disappearance of ice sheets is not a new phenomenon. It is generally believed they have been shrinking for the past 11,000 years, in fact ever since the last ice age. In Antarctica they then projected over 800 miles north of their present position.

Moreover today's rate of change is less speedy than it used to be in much of Earth's post-ice-age history. Some recent research, outlined in the January 2000 issue of *Scientific American*, provides both good and bad news. The bad states that West Antarctica's ice sheet (about the size of Mexico) may continue to shrink whether or not humans curb their production of Earth-warming gases. The good suggests that this melting ice will add to sea levels sufficiently slowly for humans 'to move their cities out of harm's way'. Such good news is therefore not tremendously good – blithely presenting an awesome task for our descendants as all the world's ports have been constructed within easy reach of the sea.

Alaskan Inuit have already been voicing unhappiness. 'About 15 years ago it started getting warmer,' said Benjamin Pungowiyi from the village of Savoonga (on St Lawrence Island off Alaska). The ice in his vicinity has been freezing later each autumn and breaking up sooner in springtime. As for the tundra that 'is not as spongy' as it used to be. Polar bears have been suffering. A study by the Canadian Wildlife Service recorded a decline in bear weight among all ages, plus a drop in youngster survival rate from 75% to 50% since the early 1980s. Black guillemots gained initially from the increased warmth, nesting further and further north in Alaska from the late 1960s. They need at least 80 days without snow to raise their young, and the birds profited from its increasing lack. Then, and since 1990, their numbers have dropped, possibly because they feed on cod which live by the disappearing ice floes.

Glaciers have been on the retreat almost everywhere. In the European Alps they have diminished by 25% in the past century. ICSI, the International Commission on Snow and Ice, predicts that most of the Himalayan glaciers will vanish within the next 40 years, unleashing a huge quantity of water. Then, in contrary fashion, that surge will be followed by a great water shortage in all the downstream communities, these having been dependent upon a steady supply, notably and all-importantly in the heat of summertime. Glacial meltwater currently creates most of the flow of some hugely important Asian rivers, such as the Ganges and Brahmaputra on which millions depend for their livelihoods.

Himalayan glaciers form the largest body of ice away from the polar regions, with thousands of cubic miles covering 17% of the Himalayan chain. Their gradual disappearance will be insidious, even if at the rate of 100 feet a year, but they can often create meltwater lakes while diminishing. These can then burst through their moraines, and have on occasion already done so. In August 1985 a breakout near the mammoth Khumbu glacier hurtled down a valley for over 50 miles, drowning numerous people and destroying a hydroelectric station. Other lakes, holding back millions of tons of water, are waiting – as it were – in the wings, with no one able to predict the looming disaster, save that summertime is more likely.

Changing sea level is a certain consequence of melting ice, but that calculated increase of two millimetres a year can seem of little concern. Even after ten years it means less than one inch. It can also seem of modest significance when compared with other recent changes, such as the world's oceans being three to six feet lower than today's level a mere 4,000 years ago. The rate of rise is thought to have been between three and six feet per century from 8000 BC to 5000 BC. The current alteration of two millimetres per year is less than eight inches per century. Overall, and since the last ice age, which finished 11,000 years ago, the global rise in sea level has been about 300 feet. (It would only have been 73 feet at today's global rate of change.) Britain was still part of continental Europe 11,000 years ago, although long since disconnected from Ireland due to the Irish Sea's considerable depth. For that reason, despite St Patrick's alleged personal role in keeping the Emerald Isle free from snakes, the serpents advancing with the warmth from continental Europe could reach the British mainland but not carry on to cross the Irish Sea.

It is also difficult to be concerned about a six-millimetre (0.23-inch) annual rise in sea level, the figure often suggested for the waters around the British Isles, allegedly rising faster than elsewhere. Not only is the twice-daily tidal range 20 feet or more, but there are storm surges which can boost the customary high-tide maximum by seven feet, literally overnight. One did so, damagingly and memorably, on that much-remembered occasion of 31 January 1953, affecting Holland most grievously of all. The surface of the United Kingdom as a whole is also tipping, with some estimates indicating that London will be 13 inches lower before the end of the 21st century. This lowering has less to do with global warming than with shifting continental plates. Finally there is a profit and loss account, with some shorelines gaining, some losing, and some relatively stable.

Around Britain this trio is evenly matched, with about a third advancing, a third retreating and a third static – for the time being. This perpetual change, with today's coastline a shape never before experienced and never to be achieved again, makes the few annual millimetres of change yet more difficult to appreciate as significant. The Roman galleys could not beach today where they started their invasion of Britain less than 20 centuries ago because that landing place is no longer attainable by sea. The Island of Thanet, today a firm and united part of Kent, was a useful and isolated springboard for post-Roman invaders, but it is no longer an island. Harlech castle, famous bastion in Wales first of English and later of Welsh defenders, used to be provisioned by boats, but no such vessels can now reach nearer than a mile. The Cinque port of Old Winchelsea, having been drowned by advancing seas, caused Edward I to build New Winchelsea in the 13th century, but the port he commanded to be constructed is also no longer serviceable. As for the English Channel, that is still widening by some 15 inches a year. So what price a few millimetres' more sea-water from one year to the next?

Accretion of new land, as with several stretches of north-eastern England, is less forthright and conspicuous than denudation, as with much of East Anglia. The slow build-up of silt and sand is bound to be less dramatic than houses tumbling into the sea. The gradual pointlessness of a harbour, with boats needing increasingly shallow draught to gain access, cannot be compared with the abrupt loss of an entire building one stormy night. The home's departure is likely to be sudden because a gale will probably serve as *coup de grâce*, not so much nibbling piecemeal at the coastline as taking a sudden bite. Conversely sand dunes do not grow in abrupt fashion, any more than silt accumulates hastily. Both take their time, much as the inward indentation of the Wash between Lincolnshire and Norfolk is getting drier, year by year. Nearly all of its deposited material is being brought down from the more northerly region between Bridlington and Spurn Head, where soft glacial clay lies on top of hard chalk. The sea there is surging inland at several feet a year, having been doing so for a long time. Dunwich – deep water harbour – in Suffolk is but a remnant of its earlier self when this community was sufficiently important to send a member to Parliament. Most of it now lies beneath the waves, with the belfry's bell allegedly tolling when the tides are right. Other much depleted regions occur on the Isle of Wight. The villagers of Blackgang on that island's south-east frontier with the sea have had to retreat 1,000 feet during the 20th century.

Inevitably, with so much of Britain so crowded, and with such extensive use made of its coastline, there is always regret when a piece of it vanishes, or threatens to vanish. Every single piece of land is either farm, or wildlife reserve, or someone's precious home and garden, or merely a treasured piece of nothing very much. Certainly every scrap of it is owned. The UK's annual turnover by the 'marine related sector of the economy', according to a major survey of coastal management published in the *Geographical Journal* in 1998, 'has been estimated at £51.2 billion'. Much is involved, such as: fisheries, harbours, industry, waste disposal, homes and homely amenities, agriculture, nature conservation, tourism, recreation. British people who holiday in Britain spend 70% of their time by the seaside. They, in general, consider the inter-tidal zone a form of birthright, even though – via distant enactments – they should be using it only 'for the purposes of navigation or fishing'. All residents and visitors are resentful of every form of loss, of well-remembered borderlines between the water and the land.

As example there is the continuing problem of Birling Gap, a splendid portion of Rudyard Kipling's Sussex by the sea. Its cliff is being eroded by some three feet a year. The surface land at the cliff's summit is owned by the National Trust, a preservation organisation particularly enamoured of its seaside properties. 'Save the Gap. Why purchase land, then let it fall in the sea?' state placards criticising the NT owners. 'Why spend £300,000,' replies the Trust, 'on constructing a rock revetment to curtail erosion – for a while?' As third force in this argument the conservationists cry out: 'Why build such artificiality and upset the natural landscape?' Slogans of 'Let nature take its course' are therefore opposed in similar numbers by 'Do nothing of the kind'. The discussion will rage, but the sea will win in the end. Its water, still widening the English Channel and eating away at the bordering cliffs, has been destructively active ever since creating this Anglo-French divide seven millennia ago.

Similar controversy – of doing something or nothing – is associated with lagoons, reed beds, wetlands in general, and every coastal nature reserve threatened with inundation. 'There must be no question of these habitats being lost,' said Britain's countryside minister in the late 1990s. Not everyone agrees, however much the losses may be regretted. To erect higher and higher sea walls is expensive, easily consuming many millions and destroying all the naturalness which embraced such wet and distinctive regions in the first place. In theory, if the sea is invading, it will create additional wetlands in advance of those it is now drowning. In practice, this may not happen in a neat give and take. Conservationists and legislators should therefore plan for the inevitable, says Britain's Meteorological Office, this organisation being more aware than most that waters will rise, perhaps for several hundred years. What is needed, says the majority opinion, is some form of strategic withdrawal, this policy sounding preferable, as with the military, to retreat. Another welcomed description is resilience rather than resistance, namely adaptation to a changing situation rather than a refusal to accept it.

It is disturbing to observe major inroads by the sea but, if nearby surface land is flat, the few millimetres of annual increase in sea level can have a disproportionate effect. Several feet of that surface can vanish each year. The incremental gains elsewhere, of more sand dunes and of more sand slowly becoming earth, can seem inadequate as palliatives. No one knows whether the total acreage of the British Isles will be more, or less, in a hundred years' time. But it is known, emotionally and disturbingly, that many relished bits of coastline, favoured by birds and by wildness in general, are vanishing, just like Dunwich which – as if in sympathy – is allegedly sounding its mournful bell. One way and another it is tolling for most of us, and certainly for those who treasure or inhabit vulnerable bits of land. (*Changing Coastlines*, by Judith Peeters, is a good and extremely factual survey of the topic.)

British concern can appear insignificant when this nation's losses are compared with the vanishing of land in other regions. According to John Houghton, former Chief Executive of the UK Meteorological Office, 'half of humanity inhabits the coastal zones around the world', with Bangladesh particularly susceptible to any rise in sea level. About 7% of its land, lived on by six million people, is less than three feet above high water. About 25% of it, lived on by 30 million, is only three times as high. As estimates have been made that sea level will soon rise in the area by one metre, a touch more than three feet, it is therefore bound to inundate at least 7% – 3,000 square miles – of Bangladesh. Global warming is blamed for about 30% of this rise, the rest not only caused by natural subsidence but the human removal of groundwater.

As 'half' of this single nation of Bangladesh was recently flooded, when a cyclone caused considerable distress, and as that country's population increase is some 3% a year, it is difficult to be optimistic about Bangladesh's future welfare, whether or not sea-level rise will add to the wretchedness. Some 85% of its people depend upon agriculture for their livelihoods. If they are displaced by rising water, coupled with savage

typhoons and population pressure, there is nowhere else for them to go, no
upcountry region waiting for development, no empty areas. There is also
no possibility of creating barriers against the incursions. Essentially the
southern half of Bangladesh is one huge delta, a filigree of waterways. The
northern half is not much better, being low-lying and only slightly less
beset with channels, dykes and streams. Mozambique is similarly beset
with low-lying land, particularly near the Limpopo's union with the sea.
Two cyclones, dropping quantities of rain well above the normal average,
caused awesome flooding there in February 2000, leading to much loss of
life and a vigorous (if belated) rescue effort by many nations.

Storm surges are, in the main, responsible for Bangladesh's massive
floodings. Some 250,000 people died during the terrifying inundation of
November 1970, and another 100,000 in April 1991. Such colossal
tragedies can be forgotten, or casually neglected, by the Western world, a
world more involved with and interested in its own mishaps. For example,
the 'horrific' North Atlantic hurricane of September 1999, named by
forecasters as Floyd and predicted as the 'greatest of the century', was
dramatically portrayed with alarming pictures as it headed for the south-
eastern United States. The coverage was intense, and forecasts were
alarming. They caused two million individuals to flee inland by car,
notably in Florida and then the Carolinas. The threat of disaster even shut
down the Disney empire near Orlando. When Floyd eventually reached
mainland America it killed four people. The migration inland had been
sensible, and probably saved lives, but the global publicity accorded to that
event-to-be was out of all proportion either to the damage it inflicted or to
any of the truly awful happenings which occur so frequently almost
anywhere in south-east Asia.

At least some of Bangladesh, the marginally higher portions, will not be
flooded in the foreseeable future. Not so with many of the myriad
archipelagos around the world, like coral islands, and like the north-south
chain of Laccadives and Maldives residing in the Indian Ocean to the
south-west of southern India. Some of the lesser islands, modest
projections above the surface, will certainly be lost. The Maldives alone
possess 1,190 individual islands, and the 34 Marshall Islands in the Pacific
are similar, with most of them no higher than ten feet above the all-
encircling sea. If there is to be a three-foot rise in their surrounding water
there would be 10–30% shoreline retreat, this leading to a land loss of 160
acres out of the 500-acre total for the Marshall group's most densely
populated island.

Flooding would certainly increase, caused by what is known as 'wave run-up and overtopping', and this would affect, or devastate, half of this most crowded atoll of them all. Much arable land would certainly be lost, with the watery invasion causing more food to be imported and a reduction in the island's availability of fresh water. As for protecting any of the atoll group from invasive and destructive sea-water the price of an effective barrier would be fearsome, costing perhaps four times as much as the gross domestic product of each piece of land. Overall it has been estimated that 150 million people will become 'environmental refugees' in the next half-century, driven from their living space by rising water, such as the 1 million on small islands, the 5 million in India, 14 million in Egypt, 15 million in Bangladesh and 30 million in China.

Sea levels are rising partly from melting ice but mainly from increased heat. The expansion of water due to rising temperature differs according to the temperature of that water in the first place. A rise of one degree C. for water at five degrees C. causes a volume increase of about one part in 10,000. Ten thousand cubic miles of water would therefore expand by one cubic mile. A similar rise of one degree C. for water at 25 degrees C. would treble the effect, and 10,000 cubic miles would therefore swell by three cubic miles. It takes time for an entire ocean, with its millions of cubic miles of water, to heat up, but even heating the top layer leads to significant expansion.

John Houghton, in his book *Global Warming*, has calculated that there will be a sea-level rise of three cms if the top 100 metres is heated from 25 to 26 degrees, namely one inch for the top 300 feet. Water does obey rules, expanding precisely according to temperature, but forecasters of future ocean levels have to make their estimates by assessing the varying temperatures of surface water, by guessing the speed at which any increase in temperature seeps into the deeper water, and by knowing of ocean currents which affect the distribution of any gain in heat. A temperature of 25 degrees is the sort of heat found in tropical waters, and temperatures of five degrees are often encountered in polar regions, but all the intervening heats at all the different levels make calculations about expansion extremely tricky, to say the least. Increased global warmth will assuredly cause water to expand, but by how much and how speedily is for the cleverest of computer models to provide better probabilities than the cleverest of humans can achieve.

Conflict Disagreement may form an agreeable basis for scientific debate, but it is less satisfactory in the public arena. In 1998 a group of scientists claimed that one effect of global warming was a cooling of the atmosphere.

A second group then asserted that the relevant data, recorded from satellites, was false because the satellites were slipping from their orbits. So which statement can be believed? The US Congress is battered, along with the rest of us, by similar forms of contradiction. The Union of Concerned Scientists, a group of 57 from 24 states, said the US government should not be distracted by the 'contrarian' views of a handful who disputed the 'mainstream scientific consensus on climate change'. 'It doesn't make sense to say we need to give equal time to the one or two per cent of scientists who are saying something totally different,' added a Democrat congressman.

Babel is defined as a confused sound of voices, with nothing to be heard above the hubbub. There is much of it today. Conflict is everywhere, and single voices are promptly drowned by countless clarion calls. Sea level will rise 'by 2 mms a year'. It will rise 'by 20 metres' should the Antarctic and Greenland ice sheets thaw. The consumption of energy is the 'number one insult' to our globe. Climate models which served as the basis for most of this concern 'were wrong'. Why promote energy efficiency when 'the market system will do that anyway'? The profit motive cares only for the present and disregards the future. The United States, in signing the Kyoto Protocol, committed the US to a reduction in carbon emissions, but its president has not yet sent the treaty to the Senate for ratification. Did that Kyoto signing therefore mean anything? What does mean anything? The planet 'will warm', but 'not as much as was feared'. How much was feared? When was it feared, and by whom? All such quotations, with every one of these coming from recent pronouncements, transform the old Babel tower into a Trappist haven of consummate peace.

Making sense of the bedlam is therefore tricky, and only modestly rewarding. 'The temperature of air at the Earth's surface has risen during the past century, but the fraction of the warming that can be attributed to anthropogenic greenhouse gases remains controversial.' True. 'What the public needs to hear from the scientific community is a greater sense of agreement about what is known.' Also true. 'Efforts designed to "pump up the volume" on the seriousness of geophysical problems may backfire.' Indeed they may. 'Since no one really knows the effects of human pollution on the environment, and since the signs of impending glaciation are mostly mysterious, is it not at least a possibility that we are looking the right way, but worrying about the wrong thing?' Possibly. Or looking the wrong way, but worrying about the right thing, say others.

It is difficult to discover a general truth among the learned papers on this all-important topic of weather alteration. 'Relative impacts of human-induced climate change and natural climate variability,' stated the headline of a major letter to *Nature* on 25 February 1999. This described how some UK scientists, from an assortment of universities and the Meteorological Office, chose to focus their modelling attention on two particular subjects, namely run-off from European rivers and the yield of European wheat. Both of these 'indicators', although seemingly unrelated, are affected by temperature and rainfall. The big question to be unravelled was whether the results of 'human-induced climate change' – via greenhouse gas emissions – could be distinguished from those due to natural climate variability. The result, eagerly awaited by this time, was that 'for some regions, the impacts of human-induced climate change by 2050 will be undetectable relative to those due to natural multi-decadal climate variability'. So there!

No one is denying that humans are influential. It is the degree of their influence that is being questioned. No one is denying that carbon dioxide is being produced by humans, or that CO_2 levels are rising, or that greenhouse gases affect climate, but human activities have to be set against the planet's own bewilderment of causes and effects. This subject has almost become a political debate – Are you with the party believing in human mismanagement, or are you not? Are you with them or us? Objective questioning is no longer the order of the day. Bruce Babbit, US Interior Secretary, instructed 3,000 scientists of the Ecological Society of America about their 'civic obligation'. It was up to them 'to convince the public' of the case for man-made global warming. 'We can't get the message through by speeches from people like me; it's all of you who have that obligation.'

The net, so available to all who wish either to receive or to transmit, has become a Babel on its own. Those against the prevailing view – that humans are harmful to the climate – can set up soapboxes even more easily than at Speakers' Corner near Marble Arch in London. The Center for the Study of Carbon Dioxide and Global Change proclaims that CO_2 is beneficial via its encouragement of plant growth. The World Climate Report, sponsored by the Greening Earth Society, is happily warm about coal, adding that most climate models fail 'reality checks'. One particular Tasmanian on the web is persistently adamant that increased snow or snow retreat is irrelevant to global warming. *New Scientist* concluded, having surfed awhile, that greenhouse sceptics 'have a taste for the blood of climatologists, for greens, and any politician to the left of Genghis Khan'.

In all of today's forums every argument elicits counter-argument. On one side is conviction that most climate change can be accounted for by naturally occurring cycles, such as ENSO, NAO and the rest. Therefore human activity can be disregarded. But, say the other side, the apparently natural cycles are perhaps triggered by human tinkering, with humans at the root of everything. The debate is a game of tennis – hit fast and a reply will come back fast, or faster still.

The debate also enters personal lives. The suggested changes – 5% less emission with a better car, the recycling of clothes, the reverential placing of differently coloured bottles in different bins – can appear inconsequential. Why economise on such pathetic issues when the planet itself is, allegedly, at stake, and when the rise in CO_2 is so inexorable? People do adjust, quite willingly, notably if the adjustment is not too hurtful. They can also behave contrarily. Following the oil price rise of 1973 many 'economised' by buying a second but smaller car, thus boosting rather than lessening consumerism. The thrust of so much individual human life is to gain more, to do better, to achieve, to triumph, and to regard family, friends and self as number one. It is not easy changing gear, moving from comprehensible personal priorities to the uncertainty of planetary concern.

Besides, why curb/curtail/diminish carbon dioxide effluent by a few per cent when 90 million extra consumers are swelling the population every year, each adding to the consumption? Why care locally when other people, and other nations, are greater villains in the planet's greenhousing? An individual sacrifice can seem immaterial to the planet as a whole, just as a personal extravagance can seem irrelevant. It is like the vote. Why bother to contribute on polling day when one ballot paper never makes a difference, with its power but a single point up from zero?

As for the planet as a whole, to what extent will Kyoto alter anything? Will the US, and others, obey their delegates? The curbings of CO_2, proposed and agreed, will not cancel the emissions of such gases. For some countries it was only 'suggested' that their rate of acceleration should be lessened, much like asking a drunkard to slow down his *increase* in consumption. Or, as with other countries, to consume no more than the considerable quantity already being imbibed. Or, with certain places, actually lessening that quantity. In no case is the drinker being asked to give up the habit altogether. That, with regard to carbon dioxide, is simply not possible. The burning of hydrocarbons is too entrenched to be abandoned, as if it is no more than a luxury item ripe for cancellation. Developed countries might cut down on car travel, on heating, on manufacture, on everything that has led to their development, but they are not about to forsake the form of living they have created.

The Intergovernmental Panel on Climate Change (IPCC), set up in 1988 as immediate response to the explosion of anxiety, has investigated possible scenarios. What if the increasing population slows down its growth, or speeds it up, or stays as it is? What about economic growth? Will it grow and, if so, at what speed? What if ocean levels do rise, if temperatures do increase, and if CO_2 levels change dramatically? Only in December 1999 was Terra launched, the first of ten satellites which have been designed to monitor the effects of human activity on the environment down below. This huge piece of equipment, sent into a polar orbit, will scan the entire planet every one to two days. Plainly it is high time, at the start of the 21st century, that such monitoring does take a good look at man-made alterations, which have been largely brought about during the 20th century – and since artificial satellites have been flown.

A particularly important climate model used by IPCC has investigated the Kyoto declarations. What would be the difference if nothing was done in response to all those affirmations or, conversely, if industrialised countries did actually adopt 15% cuts in their CO_2 emissions? The panel's computers answered that doing nothing would lead to a 1.9-degree C. rise in temperature by 2100. If the 15% promises were kept, causing much ingenuity and certain sacrifices, the temperature rise within a century would be 1.65 degrees. Therefore Kyoto, for all its trumpetings and its laudable responses to a worsening climatic situation, would adjust the global temperature increase by 0.25 of a degree within 100 years.

'Those guys weren't paying attention to their geology,' said a visitor on returning from Pompeii. Current guys are now paying attention to their climatology, but somewhat erratically. Computer predictions from the early 1980s are currently derided, the machines then being incapable of handling the huge quantities of information delivered to them. Twenty years later the computers are greatly superior, but their predictions are still regarded quizzically, mainly because so many different answers are being produced. Will scientists 20 years from now mock today's prognostications? It is highly probable. With computer advancement occurring at such a pace, far faster even than before, it is easy to believe that now will be a kind of mediaeval age when viewed from 2020.

The computers will certainly alter, and so too their forecasts, but human reluctance to change its ways may stay identical, being the single probability in a sea of variation. In 1998 John Ashcroft, of Missouri, attached a Republican amendment to a bill which would ban the administration from spending any money on 'rules, regulations, or programs designed to implement, or in contemplation of implementing' the Kyoto agreement. 'Why should I do anything for posterity?' says the old adage, and so say a mountain of individuals now sharing planet Earth. Or perhaps they will actually concur in slowing down their rate of increase in consuming things, if that does not sound too bad. Continue to grow, but less speedily. Be human, but a touch more knowingly. Realise that all of us have but a single home.

Regarding the past when contemplating the future can make one wonder about our ancestors. We, after all, are merely them but in a different guise. They did daft things, and terrible and cruel things, and went to war with the greatest of ease. They were short-sighted, inept, unthinking. Why therefore should we, their genetic descendants, be any different? How was it possible for World War One to consume so readily such a quantity of life, with soldiers running through mud and wire at machine-guns? Why all those religious wars between variants of Christianity? How does dogma gain such a hold, again and again, with logic vanished absolutely? It is so easy to despair of earlier convictions, earlier aspirations, and all earlier wrongfulness. Each new generation may have the conceit to believe it is somehow different, even superior, but the facts of history belie such wishfulness. It is even arguable that stupidity is on the increase, with the 20th century witness to greater awfulness than any of its predecessors. How could it be that around 55 million lost their lives in the maelstrom of World War Two? The whole planet, more or less, had gone to war.

As of the new millennium it is easy to wonder if a new age will be accompanied by a greater understanding of universal needs. Somehow we six billion people must appreciate, as the next chapter relates, our lonely and united situation within a sea of emptiness. Our solitary ark is a pale blue dot encircling its sun, and – as yet – the only known one of its kind.

Chapter 16

THE VIEW AHEAD

Pale blue dot – population numbers – human fertility – life expectancy – gas predictions –
IPCC forecasts – government or corporation – technological change – hydrogen, nitrogen –
carbon disposal – another revolution?

In the dedication to my father at this book's outset I unkindly suggested
that he, along with many of us, was an unofficial member of the flat earth
society. It is so difficult, during the running of ordinary lives, to appreciate
that everything is taking place upon an 8,000-mile diameter ball revolving
in space, which is quite unsupported save by gravitational forces from
elsewhere. The fact of the bus being late, money being tight, or the entire
gamut of our lives, can all seem so trivial in comparison with our awesome
situation within a universe of unbelievable dimensions.

Even within our solar system the measurements are so vast. The sun is,
on average, 92,900,000 miles distant, with its light not reaching us for eight
minutes. The planet Pluto, another member of our solar community, only
receives its sunshine from the same source after seven hours. We exist
within the Milky Way galaxy, and our sun with its attendant planets lies
about 25,000 light years from this galaxy's centre (or 150 million billion
miles). This sun of ours, together with its solar system, is orbiting that
distant centre, and does so once every 225 million years. Our local
arrangement forms an extremely modest part of a spiral-shaped galaxy
consisting of a few hundred billion stars. Astronomers used to think, not so
long ago, that the Milky Way was the entire universe. They now know that
at least 100 billion other galaxies are out there. In short, roll up, roll up, to
join the flat earth society. Doing so can seem to make more sense than
attempting to comprehend the bewilderments on offer from today's
professional star-gazers.

Genesis states that God made the firmament on the second day and
'called the firmament heaven'. Is there any religion whose creator was
content solely to make a single planet, the one called Earth, this lying

within the mind-numbing vastness of a star-infested space? Carl Sagan, author of the widely read *Cosmos*, debated whether religion and astronomy could happily coexist in another book he called *Pale Blue Dot*, this title arising after Voyager had taken an inspirational photograph. Having done its work as it passed by neighbouring planets this spacecraft was speeding away from Earth at 40,000 miles an hour, being already 3.7 billion miles distant in February 1990 when it was commanded to look back and photograph the planet which had made and launched it. When this picture was pieced together, after its several weeks of transmission time, the result showed a skyful of stars, much as we can see at night when looking upwards, but there, in the middle of Voyager's photograph, was a pale blue dot. That was us. That is what we are, and all of us are there. Our planetary home is but a single speck, illuminated by its parent sun, and utterly insignificant when set within the firmament of which it is no more than a single grain of sand within a desert of incomprehensible dimensions. We are no more than one solitary planet within a galaxy of a few hundred billion stars (and no doubt great quantities of other planets), this single galactic arrangement we call the Milky Way being distantly accompanied by a hundred billion other galaxies.

Most of what happens in the universe is irrelevant to us on our planetary home. The other galaxies are of no consequence, and neither is most of our own, but it is incorrect to review our weather as if Earth is entirely independent of anything beyond its boundary. It needs to be set in context, and that immediately involves the sun. This has cycles, which are unerringly influential, and its radiation comes to us in different forms. The moon, our single satellite, is also relevant. So too the comets which orbit our sun, perhaps in a few dozen years like Halley's, or perhaps in several thousand years, or even longer. Their timings are their own, but they can assault us, as (probably) happened at Tunguska, or merely pass us by, making a spectacle for wonder. Rather less wondrous are the meteorites which can lay waste and devastate. They have the power, if they are big enough, to extinguish every living thing, having very nearly done so on occasions in the past.

When Earth is regarded as a spherical object, like the other planets encircling our sun, its climatic happenings begin to make some sense. All of it is warmed by solar radiation, but unevenly owing to its rounded shape. The tilt of its axis adds to that unevenness, creating summers and winters, the annual darknesses and brightnesses which would not occur in similar fashion if that tilt did not exist. There are also the earthly revolutions, which lead to our nights and days. 'And the evening and the morning were the first day,' this biblical creation taking place even before the firmament. As for the third day's labour, the separation of land and sea, that distinction is also best realised from the sun's perspective, with its warmth having such different effects whether reaching water or dry land.

I unkindly likened my father to an ant seeing its territory as if that was all, as if landscapes can exist in isolation. Instead everything and everywhere is interdependent. The weather above Dorset, the swallows' arrival time, the lengthening and warming of the days are caused by other circumstances. This planet of ours may be lonely and unattached, but it is subject to extraneous influence by being where it is. Hence my wish to embrace the topic of weather only after discussing the astronomical imperatives. Hence also the wish to talk of climate change only after detailing climate in general, the basic parameters of wind and radiation which form our weather's fundamentals. The prospect of change makes no sense unless there is earlier understanding of what is being altered. And the possibilities of man-made alteration are more comprehensible only after natural fluctuations have been explored.

So much for the intent of this book, but one crucial and additional variant remains which is both natural, and yet not so, simultaneously. Human numbers in the past have fluctuated, as with all of life, in standard fashion. Bad years killed people, and good ones permitted a regaining of their strength. There was no control of any kind, save from life's harsh truths, but no longer is this ancient ruling the guideline that it used to be. Human beings, in the main, can now decide their own fertility. As people are exerting their authority upon this planet it is therefore highly relevant whether there are to be more or less of them. Right now, as the whole world knows, there are more of them than ever before, and therefore more to damage the planet should this continue to be their form of recklessness. What the demographers have to say about our future density is vital information.

Population For those born on 12 October 1999 they can take comfort, or maybe discomfort, from the fact that one of them will have become the six billionth human to be alive on planet Earth. On that day the United Nations affirmed that this milestone had been reached, a figure never before achieved by *Homo sapiens*. The five billion mark had been reached in 1987, which meant that 228,000 extra individuals had, on average, been arriving every single day during that 12-year span. At the start of the 20th century there were fewer than two billion – the past century marked the most explosive growth in human numbers there has ever been. It may well be called the century of world wars, or the one which started aviation, computing, radio, transistors, space travel, transplants or a thousand other initiatives, but it certainly witnessed the arrival of more people. Whether more babies were born in the 20th century than in several previous centuries is not possible to say, since so many who were born in earlier times lived such little lives, but it is certain that more of them survived in the last 100 years.

As for Britain's 147-square-mile Isle of Wight being able to contain the world's population, this being a famous yardstick, that is still possible. Each individual of the six billion would have four square feet. Within the District of Columbia, home to the US capital and approximately half Wight's size, the ration per person goes down to two square feet (still better than the average city elevator or rush-hour train). If the number of Earth's people is divided into Earth's land surface there are 105 of them to each square mile – about the proportion now present in Kenya, Iraq or Mexico. It can therefore seem as if the planet is not overcrowded, save that practically everyone is packed on to land suitable for agriculture, the 12% of Earth's terrestrial surface on which useful crops can grow. To fly from one major city to another equally crowded community elsewhere in the world can make it appear as if the world is already full. That is not so as there are all the emptinesses in between, such as the six million square miles of the ten largest deserts or the five million square miles of tundra, but it is certainly fuller of human beings than it has ever previously been.

The number of people is entirely relevant to global warming, to CO_2 emissions, to rising ocean levels, and to all the matter within this book. It is also pertinent that the current high in population will become yet higher, the human density still increasing so vigorously. Birth rates have stabilised in both Europe and North America, but the developing world is increasing its population by 1.7% a year. Its nations account for 98% of the growth, and their total numbers will double in the next 40 years. As more than 400,000 successful conceptions are now taking place every day the next billion landmark will therefore topple before too long, with seven billion living on Earth a decade or so from now, and eight billion not much later. The only bright light in this particular tunnel of increasing crowdedness is that demographers think (and also hope) there will be stability before the end of the 21st century. By then we will have seen the end of population doublings. Today's six billion will therefore not become 12 billion, that is if the forecasters are right.

Population density around the world will certainly change. The United States, currently third nation in people numbers after China and India, will soon be ousted from that position by Pakistan. Nigeria is expected to become number five before too long, with Africa as a whole having the highest birth rate of any continent. Even in the past 20 years its numbers have grown from 470 million to 763 million. Some people have been asserting that, as HIV is such a killer, notably in Africa, it alone is preventing, or will prevent, population growth. That is not so. WHO reported in November 1999 that '50 million individuals worldwide' have been infected with HIV since the outbreak began and 'over 16 million have died'. (Africa is now gaining over 15 million people every year.) The AIDS deaths in 1999 reached a record 2.6 million, 'with an estimated 5.6 million adults and children worldwide' becoming HIV-infected that same year. AIDS may not be losing its momentum, but population growth is not doing so either. With people numbers, for the planet as a whole, increasing by over 80 million a year the slaughter caused by AIDS is still only 3% of that massive figure.

Assisted by immigration, which accounts for one-third of American growth, the United States will probably add 120 million to its total in the next 50 years, or 6,570 more people each day. Europe and Japan, although similar as industrial areas, are expecting to see their populations shrink. Such decreases among nations which are currently liberal in producing CO_2 may seem a blessing, but the developing countries are rightly named. In the main they *are* developing, and are therefore increasing their outputs of greenhouse gases per nation and per individual.

Official UN estimates state that the global population will reach 10.8 billion by 2050 if the number of babies per woman falls to the replacement figure with each pair of parents replicating their number. If the average number does not fall so low, and merely reaches 2.6 children per woman, the population by 2050 will be 27 billion. The arithmetic therefore changes alarmingly. Fortunately it is expected that women will only be having 2.3 babies by 2025, a far better number than 2.9 in 1995 and 5.0 in 1950, these being the average figures for the whole world.

Reduction in baby numbers would be yet greater if contraception was more widely available. Researchers with the Asian Development Bank, based in Laos, asked people what kind of help they most required. The men said jobs, while the women requested family planning. In Ethiopia only 4% of women use contraception and they have seven babies on average. In South Africa some 57% use it, and the fertility rate is 3.3. Within 'nearly every country surveyed', according to Malcolm Potts in the January 2000 issue of *Scientific American*, 'women are bearing more offspring than they intend'. Save for Europe there is a straight relationship between contraception use and fertility rate. That is, after all, what the word means. It is against conception.

Whether women have 2.6, 2.7, 2.8 or any similar average number of babies is so influential on future population numbers, each 0.1% of change having such dramatic effect, that it is important to remember the powerful influence of postponed births upon the arithmetic. In Britain the number of babies born to women in their 30s has become greater than the number born to women in their 20s, a remarkable alteration occurring for the first time in the late 1990s. (In 1970 the average age of all mothers had been 26.) If such a trend of later births were to happen elsewhere in the world, notably in the relatively prolific developing world, the result could be as beneficial in reducing global numbers as a lessening in the quantity of children born to each mother.

Currently there are only 65 nations out of the total of 192 which are either reproducing at replacement level or are having even fewer children. With the other 127 nations their fertility rates are still higher than replacement, but these are falling and have done so markedly. The average developing nation figure in 1950 of 6 babies per woman fell to 3.3 by 1998. If the world's population is not to double again it will have to drop still further, but no one is expecting it to reach replacement level just yet for the planet as a whole. Even if it did so – right now – the world's population would continue to grow owing to the great number of people in the younger age groups who either are, or are about to be, in the business of procreation. Out of the present six billion total almost half are under the age of 25, with one billion of them between 15 and 24. Their style of reproduction, and how many children they have, is therefore still ahead of them for about 50% of the world's people. What they do, and how many offspring they produce, is of huge importance to climate change. Cutting carbon emissions by 10%, and all the other proposed modifications, will have their benefits overwhelmed if millions more people arrive to trample on them.

Life expectancy will continue to improve, a fact which helps – however contrarily this may seem – to lessen population growth. If children do live longer their parents will have fewer of them. Whereas the developed nations have seen their life expectancies rise from 66 years, on average in 1950, to 75 today, the developing countries have experienced the far greater rise from 40 to 63 in the same period. Contraception, improved health and a better standard of living have all assisted in this change, with better life expectation leading to fewer children which, in turn, means a further improvement in the living standard. So many circles are said to be vicious, with each harsh fact leading on to others, but fewer children leading to better lives and then to fewer children is an opposing, and most unvicious, circular arrangement. Fertility is always reduced as life expectancy goes up. There is even said to be a cross-over point when fertility falls below replacement level. Pooled information from many sources indicates that this happens when life expectancy reaches an average span of 79.5 years.

As for the future it is thought that longevity will continue to improve, reaching an average of 81 years and 76 years in the two kinds of nation by the year 2050. Even by 2025 it is predicted that 26 countries will have a life expectancy at birth of over 80 years, with Iceland, Italy, Japan and Sweden doing best of all, their expectations being 82 years. Australia, Canada, France, Greece, Netherlands, Singapore, Spain and Switzerland come second at 81 years. Countries at the other end, such as Angola, Sierra Leone, Rwanda, Afghanistan and Malawi, will all have life spans less than 60 when 2025 arrives. Older people are not necessarily more consuming of resources, or more destructive of the environment – if anything they are less so – but their increasing existence indicates a healthier, wealthier, and (probably) more extravagant society. The number of Earth's people over 65 will rise from today's figure of 390 million to 800 million by 2025. They will then form 10% of the total population. As for the over-60s their number will grow from the current 590 million to 1.2 billion in the next quarter of a century.

An organisation known as Future Foundation predicts that British children of the years ahead, and no doubt those in similar countries, will lead extraordinarily different lives. Individuals born in 2010, according to this foundation, are unlikely to retire until reaching 80 and will live until 120. These people will have time and opportunity for three or more careers, several educational stints, a couple of families, and 40 final years of leisure. If such a forecast does come to pass it will not be universal, although longevity is increasing everywhere.

Unfortunately there is no suggestion that energetic Methuselahs will be any less destructive of the planet than the current crop of advantaged and elderly human beings. For individuals desperately concerned about the planet's welfare, and about the negative role being played by people, there is even the Voluntary Human Extinction Movement (which promotes itself on the net). Its creed is to 'phase out the human race by voluntarily ceasing to breed', thus allowing Earth's biosphere to return to good health. 'Each time someone decides not to add another one to the burgeoning billions ... another ray of light shines through the gloom.' 'Humans must go,' adds the Gaia Liberation Front, also on the net.

Gas Predictions Estimating the number of children to be born in the 21st century is, as already implied, one form of impossibility, but the demographers are at work and they express muted confidence in their estimates. Forecasting the emission of carbon, and therefore levels of greenhouse warming, is another intractable problem, this one becoming more severe with time. The IPCC, already mentioned as the international organisation set up in 1988 in prompt response to the sudden concern about climate alteration, has become much more dubious about its predictions. In essence it is now saying, via a report in March 2000, that carbon dioxide emissions might quintuple by 2100, or might fall by 30%, or achieve a figure somewhere in between. As a letter to *Nature* began in January 2000: 'The distribution of sources and sinks of carbon among the world's ecosystems is uncertain.' Indeed it is. 'Some analyses show northern mid-latitude lands to be a large sink,' the letter continued, 'whereas the tropics are a net source; other analyses show the tropics to be nearly neutral, whereas northern mid-latitudes are a small sink.'

As Franz-Josef Radermacher, of the University of Ulm, wrote in January 2000: 'It's one thing working out whether, how and why the Earth's climate is changing. It's quite another working out what to do about it.' Not only do decisions have to be taken *before* all the evidence has been gathered in order to prevent possible disasters, such as massive crop failure, but these decisions have to be taken internationally. Think of all the conflicting interests! Think of the national frontiers criss-crossing the planet, and of the single-minded nations they embrace! Or do not think of such things, and attack the problems as if Earth's people now realise they all exist on the same planet, rather than within counties, or states, or even federations. Iceland will have to agree with Bangladesh, and Brazil with Myanmar. Keeping warfare at bay, or sufficiently moderated, has been one task for the United Nations. Keeping the planet healthy, or sufficiently so for reasonable living, will be yet more onerous a labour, partly because so many uncertainties are involved.

Early estimates confidently stated that greenhouse gases would double their current levels within a century if governments did nothing to curb them. As a result the global temperature would rise by two to four degrees C. The IPCC now considers it unrealistic to assume such a negative response by governments but is at a loss to know what all the numerous responses will be. Consequently it has made various presumptions. These have been based upon population growth (which no one knows), upon technological advance (also uncertain) and economic welfare (possibly the most uncertain of them all). For an individual it is like assessing personal wealth in the future when job promotion, family size, personal health and governmental tax levels are all in doubt. Will there be cash available for three holidays a year, or one weekend, or merely bankruptcy?

Some 40 possible scenarios have been fed into the IPCC computers – more people, even more, less CO_2, more CO_2, good economies, bad ones, and numerous possibilities in between. It is thought there may be either 4.3 billion tons of CO_2 emissions by 2100 or 36.7 billion, and forecasting confidence has gone about the previous estimate of 18 billion tons. If bet-hedging is to be an aim the IPCC should surely win some kind of prize, much like weather forecasters promising dry and windless days possibly interspersed by fog, cloudiness, drizzle, rain or gale. But IPCC has a point, and is not afraid to say so. 'There can be no best guess. The future is inherently unpredictable. Views will differ on which of the scenarios could be more likely.'

There is also a belief, making prediction even harder, that administrations will be less in charge of future change than big business. Fred Pearce, *New Scientist*'s principal writer on climate issues, put his views succinctly in an 'open letter to the eco-warriors': 'We have learned one thing – governments don't count so much any more. In the emerging global market place they are the pawns. It is corporations that have the real clout.' Governments have to follow, or at least persuade, the majority of their constituents. Corporations can act the moment they see commercial advantage. 'It will be companies,' concluded Pearce, 'that will trigger the political sea change necessary for real reductions in greenhouse gas emissions.' In similar vein Myles Allen, of the Rutherford Appleton Laboratory near Oxford, wrote in *Nature* in October 1999: 'As long as politicians only propose targets that they think they can sell to a generally indifferent electorate, it seems unlikely that we will achieve the order-of-magnitude reductions in greenhouse-gas emissions required to make a significant impact on the problem.'

The trouble with global warming, as many have stated, is that it is global. The trouble with administration, as we all know, is that it is piecemeal, a jigsaw of frontiers, a mosaic of different communities each thinking differently, acting differently, and only occasionally assembling for some united effort. From nature's viewpoint, if it could have such a thing, the presence of these (usually) invisible border lines is the most bizarre aspect of all mankind's activities. Climate knows no barriers.

Neither do migrants as they wing their way across the latitudes, save that they are more likely to be shot, maimed, trapped or eaten in some areas. A visiting alien would have to learn that people east of some longitude pay four times more for fuel, or half as much in taxes, or nothing on armaments, or ten times more on them than on health. The curbs on global warming will need a global attitude.

There is precedent for this concern. The two world wars did affect the world. Even many non-belligerents had a difficult time. The Wall Street crash of 1929 did send shock waves right around the planet, much as Krakatau did. The corporations are better placed to act globally than are the governments, and the biggest ones are the internationals. Their title holds some opprobrium, embracing an attitude to profit often seen as inimical to human welfare, but the major companies do not regard the world as fragments, as awkward pieces intriguingly distinct. They have already seen the picture that all these pieces make, and their business is with that large-scale offering rather than its parts.

Governments may express official worry about the possibility of increasing hurricanes, rising sea levels, droughts, and floods. They are quick to proclaim disaster areas after tragedies have struck. They may even offer financial aid while their representatives are publicly speedy to express some grief. Insurance companies, horrified by claims sometimes running into billions, feel quite a different kind of grief. They are appalled by each disaster, knowing of its cost, and aware it could destroy their business. If there are changes to be made, new laws to diminish future catastrophes, or new technologies to be implemented, it is insurance companies and their kind who will act as goad to see these things are done. They may work with the eco-warriors, and the warriors may work with them, each seeing benefit in cooperation.

What will make us insulate our homes more effectively, the better to save on CO_2 exhaust – governmental edict or individual profit – after a way has been found to save us cash? If big business finds it cheaper to install millions of solar panels rather than searching for ocean oil it is certain that sunshine will immediately supply a higher proportion of our energy. Moreover, as those millions of panels are deployed, their price will tumble to make the switch more profitable for everyone concerned. A government could not initiate such a scheme and order such alternative solar systems to be installed, particularly if – at first – they are more expensive. The change will have to come from the much-derided market place, from people's pockets rather than a fear of CO_2.

Consuming and squandering are rife in the developed world, but did not use to be. A couple of generations ago 'being economical' was a prevailing wish. On a personal note the business of staying with elderly relatives could be nightmarish. Ice appeared on the *inside* of bedroom windows. Cold was pervasive. Passageways were in darkness. Single logs gave flickering light, suggesting but not donating heat. Urine would freeze in chamber-pots, and several sweaters were standard daytime wear. Householders today, save for the solitary and elderly, tend to bask in warmth. Thermostats are turned high. Lights are on in every room. Television is often left to run, entertaining no one. Of all the bills which must be paid the price of energy at home is modest relative to meals in restaurants, to hotel nights, to travel overseas, to the blessing of a car.

'There has to be a carbon tax,' proclaim environmentalists. That would make consumers less profligate with energy, they say, and economies would result. People will act if there is gain, if public transport becomes a better option, if there is advantage in being virtuous. What does it matter what I do, so runs a general refrain on gas emissions, when 273 additional people are arriving on the planet every single hour, each such batch of newcomers destined to produce far more carbon than one individual could ever use or save? What does it matter when all the wealthiest are determinedly culpable? (Miami's airport is so chilled in summer's heat that external jaunts are necessary to regain some warmth, a procedure every lizard would immediately comprehend.)

Technological change An inspector of inventions was famously, and remarkably, prescient in assessing the future value of all articles set before him. When asked for his secret he replied that it was simple. He turned every one of them down, and was therefore 99.9% correct. It has already been said that the 20th century witnessed the arrival of aviation, computing, radio, transistors, space travel, transplants and a thousand other initiatives, but who could have foreseen the progress that each of these have made? Who could even have foreseen all their introductions? Similarly what will happen in the 21st century, of particular relevance to climate change, now that it is here?

Oil as a fuel will have to expire, with 40 more years or so to go. What will then take its place? Coal may regain a greater share of energy production, but its side effects will be no more welcome then than now. In any case all fossil fuels have limits to their supply as, to borrow Mark Twain's phrase about the land, they are not making any more. There is speculation that hydrogen may become the future fuel. If so Iceland will be in the lead, having already taken steps in that direction. It has abundant hydropower, and will use that energy to split molecules of water to gain the gas. This will then be liquefied before being fed to fuel cells. Some major names are involved with Iceland, such as DaimlerChrysler (which has a prototype fuel cell vehicle), Shell (which has a hydrogen filling station in Hamburg), Norsk Hydro (which is big in hydro-electricity) and Ballard Power Systems (which makes fuel cells). Iceland aims to have severed all its links with fossil fuel 20 years from now.

There is also talk that nitrogen could be a future power source. It is undoubtedly abundant, and therefore cheap, being 79% of all air. It would first have to be liquefied and then used, much as steam was used, to pump pistons. The cost, according to its promoters, would be no more than today's electric vehicles, and tanks of 80 gallons could send cars 250 miles on a single filling. One extra advantage, apart from being non-polluting, would be that any carbon dioxide in air being frozen to gain the nitrogen would solidify long before nitrogen's freezing point, and could therefore be discarded. Nitrogen also liquefies well above hydrogen's temperature, a further gain. Frostbite might result during accidents, an alternative to the current possibility of incineration.

The future might also pay more attention to the sea. Ocean water has been experimentally fertilised with, for example, iron sulphate, causing plankton 'blooms' and therefore a massive absorption of carbon dioxide. Fish could be fed in a similar fashion, and harvests would be multiplied. Humboldt's famous current, which has its cold temperatures switched off in El Niño years, is so bountiful for anchovies, in particular, only because it scours up nutrients from the depths. Something similar could be created artificially. In general terms Neolithic Man would feel quite at home with most modern food production. We grow crops in fields, much as he learned to do, and we suffer when weather is unhelpful, just as he did. We may hunt for fish with better nets, better boats and sonar, but the process is still a form of hunting which he would understand. The 21st century may well see alteration of such neolithic ways. It may be forced to do so.

As for the rising levels of CO_2 there are numerous suggestions for reducing them, quite apart from lessening that gas's production in the first place. Injecting the gas directly into the depths of oceans might help. So too would the encouragement of all forms of plant growth, and not just reforestation. Stirring up silt from deltas can encourage algal growth at sea. Using nickel-based turbine blades, as in jet engines, instead of steel blades could raise power station operating temperatures, making combustion more efficient. Landfill sites, so unsightly and so indicative of modern wastefulness, could be regarded more positively as splendid burial chambers, with the carbon of wood and paper banished beneath the ground. The United States is believed to lock away some 30 million tons of carbon a year. Even termites have been subjected to scrutiny. Altering their gut bacteria, a practical possibility, could lower their output of methane, a greenhouse gas more powerful than CO_2. More research could be dedicated towards what is being called the ignorosphere, the atmosphere's relatively unknown region higher than ten miles above the Earth. How do greenhouse gases affect it, and what effect does it have on greenhouse gases? Not only does this region contain quite a bit of the atmosphere's mass, but also the ozone layer which is so protective against UV radiation.

A further problem for the future concerns the reaction of human beings to – well – to everything. Economists are steadfastly aware of the difficulty. When people, by rights and according to the rules, ought to be holding their money tight they suddenly indulge in a spending spree. They save when they should spend, and do not always buy if things are cheap. So too with energy and its conservation. If there is official policy, and even stricture, to insulate roofs, install double glazing and curtain windows to save on fuel bills, there will be extra money for householders to spend – on bigger boilers, larger refrigerators, extra heat in winter and extra air conditioning in summer. (It is possible for visitors to believe that North American homes are often warmer in winter than in summer. Would one's relations from an earlier age, with their ice-encrusted window frames and visitor urine frozen in the pot, perish in such conditions?)

Will the 21st century actually witness a revolution rather than trifling amendments here and there? Worker productivity expanded 200-fold when coal was used instead of human muscle. It is therefore right to call that industrial switch a revolution. Materials used by industry's metabolism in the US every day amount to 20 times the weight of the American population, according to Norman Myers, already quoted in this book for robust opinion concerning conservation. More worryingly he adds that 'only one per cent of these materials ends up in products that are still in use six months after being sold, the rest being junked'. He suggests 'industrial ecology', with each manufacturer feeding upon the wastes of other manufacturers until emissions are finally reduced to zero. This, after all, is a well-tried and tested system, with years and years of satisfactory experience. It is what nature does.

'Nothing dates as quickly as a prediction' stated a *Nature* editorial, after pronouncing that it was asking various individuals to think about the scientific advances of the millennium to come. Arthur C. Clarke, the first such author, and one with an impressive list of correct forecasts behind him, such as geo-stationary satellites, was depressingly dismissive about humans of the future when writing a fictional report about their involvement in a future century. 'By mishandling the ultimate forces of the Universe, they triggered a cataclysm which detonated their planet.'

A different method of prediction, also depressing in its way, is to look backwards and see if there is precedent for the current situation. Unfortunately there is. Ice cores from arctic regions show a close correlation between carbon dioxide levels and ice ages. When CO_2 is high the planet Earth goes cold. Right now, for all the talk of global warming, these levels are very high. In fact they are higher than at many cold times in the past. And, which is even more to the point, they are getting higher every day. It is therefore entirely proper that the world, the entire world, is beginning to think about its global problems, its warming, its CO_2, the total sum of all its individual actions. Never before has it done so. It is high time that it did.

Finally, as the brilliant nuclear physicist, Niels Bohr, once said: 'Prediction is very difficult.' He then thought a while before adding: 'Especially about the future.'

EPILOGUE

'The problem about the future,' stated a *New Scientist* editorial in July 1999, 'is that it keeps arriving far too early.' That magazine, in addressing the ethical issues caused by biological advance, could well have been expressing concern about almost any aspect of its interests, not least the changing climate. To reiterate the sentiment expressed in this book's opening section, how amazing that a wide-ranging book called *The Seasons*, published only 30 years ago, contained 'not the smallest mention of global warming, of ozone depletion, carbon dioxide quantities, methane, glacial retreat, El Niño, La Niña, rising sea levels, low-lying inundation or greenhouse conditions'. It is easier to wonder what it did mention, if disregarding all those topics, it being a large book about the vagaries of climate.

Prediction, as Niels Bohr might also have said, is very difficult even about the past. The 20th century must have witnessed more change than any of its predecessors, but which of the countless alterations was – and will be – most wide-ranging in its effects? Nuclear power, both peacefully and horrifyingly, has already been an awesome change. So too aircraft which are still less than a century old, and radio and television which are younger still. As for computers their rate of change is still phenomenal, each new development making even a previous decade quite archaic. Space flight has been with us for less than half a century, with nothing rocketed into orbit until 1957, no human circling the Earth until 1961, and no one landing on the moon until 1969.

But, according to some, none of this has been as consequential for humankind as the Haber-Bosch process for synthesising ammonia. Without this invention the world's population could not possibly have grown from the 1.6 billion it was in 1900 to the 6 billion it is today. The date of 3 July 1909 should therefore be as well known, if not better known, than the Wrights' first flight of 17 December 1903 or the first nuclear chain reaction of 2 December 1942. In that summer of 1909 a large apparatus for producing liquid ammonia functioned perfectly for the very first time.

Commercial production began on 9 September 1913, just 50 months after the laboratory demonstration.

Global output of ammonia is now 130 million tons a year, of which 80% goes into fertilisers. Without the Haber-Bosch process, and without all its patented descendants, many involving urea, it has been calculated that almost two-fifths of the world's population would not now exist. Yet more importantly the billions to come would also not exist. Whatever one's opinion may be about the virtue, or otherwise, of so many more mouths to feed the fact of their current and impending arrival is undoubtedly of major consequence. Nuclear power, aircraft, antibiotics and television can quickly pale into insignificance, with the massive increase in population known to exacerbate every problem. Such as pollution and deforestation. Such as rising levels of greenhouse gases, and such as global warming.

Whether or not humans are to blame, and yet more blameworthy for continuing to proliferate, the fact remains that 1999 was a very hot year. Within the United Kingdom it was quite the hottest that has ever been recorded since 1659 when measurements (of any climatic kind) were first made. It was not the sunniest there has ever been, that honour resting with 1995. Nevertheless both dates suggest that temperature change is happening, and the first decade of the 21st century is likely to witness climatic records tumbling yet again. February 2000 was in Britain the eighth month in succession which had been warmer than normal.

The century of the 1900s was also wetter than had been expected, as the global average annual rainfall increased by 22 millimetres – almost one inch – within those hundred years. More specifically, and over North America, northern Europe, Asia, Argentina and Australia, it increased by 40 millimetres (1.5 inches). There is no doubt that change is happening, and the latter part of the 20th century will become known as the period when many people – and even some in authority – woke up to the fact. The earliest part of the present 21st century will therefore be known as the period when people either did, or did not, do something to arrest the trend.

There has already been some change in official circles. After the 1990s had been recorded as the warmest decade of the past millennium the attitude of the Intergovernmental Panel on Climate Change adjusted its thinking. In its 1994 report, second in the series, it had considered only that the 'balance of evidence suggested a discernible human influence on global climate'. In its third such report, due for publication in 2001 (which was circulated prematurely for comments), it lays the blame squarely upon humans and their activities. There is no longer doubt or hesitation. Instead there is certainty. Had humans not exerted their various forms of influence, say the IPCC experts, the planet would actually have cooled in recent decades. Instead, and unequivocally, our Earth has warmed.

New Scientist published a trenchant editorial on this change of heart in November 1999. 'With the help of increasingly successful climate models [the IPCC] lays out our likely future in awful detail. We have probably already signed death warrants for several low-lying Pacific islands, casualties of rising sea levels. And if we let concentrations of greenhouse gases increase more than 50% above present levels – which could happen by the middle of the 21st century – the Amazon rainforest will simply shrivel up and die.' There are still numerous uncertainties, such as the 'alarming fuzziness about atmospheric mechanisms that could turn small changes in solar radiation into large temperature swings here on Earth', but these, it is argued, should be cause for more concern and not an excuse for delaying action. The 21st century has therefore much to do. 'It is now time,' concludes the editorial, 'for governments to show that they can act as one to halt the coming nightmare.'

At this stage in this book, with many a glance forwards as well as backwards, I wonder what will be missing from its index which will seem a strange omission when regarded from the future. A crystal ball would be fun – perhaps, and would certainly be useful. At least it might show that the great Arthur C. Clarke was wrong for once (in presuming our self-destruction). Naturally, along with six billion others, I sincerely hope he is, and that we are here to stay; but, when looking back at the mayhem made by people, and mere mismanagement, it is possible to think that *Homo sapiens* is not much entitled to continuing existence. Is there any other species quite so guilty and so undeserving of a future role?

On the other hand I would hate for my children, and those I love, not to have the prospect of future generations endlessly ahead of them, these initially embracing their children and all those whom these further children love. That is the strength and the weakness of the human condition. We are each selfish within our personal environments. We do look after our own better than those who are elsewhere. Somehow or other this narrow and personal focus will have to encompass everything about us, and that includes the planet. There is no point in having a decent home, as Thoreau said, unless you have a tolerable planet to put it on. This Earth of ours has been given a rough time, particularly in recent years, with our pale blue dot richly deserving far kinder treatment.

What better time is there to start upon this planetary welfare than right now? There is, so far as we know, nothing else quite like this Earth in the whole astonishing universe. It is a beautiful dot, dwarfed by all the distances on every side, and home to a species with the power to make it an even better, more habitable and more enchanting place – or something infinitely worse.

The problem is a little like life itself. We each know we will die, and yet operate – with trivial daily concerns – as if that time will never come. Similarly we know our remorseless consumption of fossil fuel is doing harm, will continue to do harm, and will be harmful even if we halted CO_2 emissions right now. Thoughts of an unwelcome future do not occupy our minds. We speak of 10% cuts here, and a little more renewable energy there, but are much like householders polishing the furniture when a lethal hurricane is on its way. Save the wetlands, save a precious species, save bits and pieces, state the conservationists and, if it is not too inconvenient or distracting, spare a thought for the planet itself.

Of course, and in addition, there is the gross possibility that a meteorite will be horrendously damaging. There is also the chance that some super-volcano will blow its top and be equally disruptive. But, as counterweight, there is the certainty that Planet Earth is already warming, will warm further, and will experience in consequence all manner of upheaval. Whether we inhabitants will suffer more from the extremities of weather, from rising sea levels, from shifting patterns of disease, from switched ocean currents, from drought in place of rain, storm in place of drought, or human resentment and possible warfare from increased hurt, is not yet known. The only surety is that damaging change is on the way. As seen from the start of this new millennium, with all the relatively trifling worries which dominate and assault us daily, the suspicion surfaces that we will – probably – do little about it.

What we should be doing is a very great deal.

INDEX